Magic Mushrooms

by Peter Stafford

Ronin Publishing

Berkeley, CA

Magic Mushrooms

Copyright 2003 by Peter Stafford

Derived from Psychedelics Encyclopedia
Copyright 1978, 1983, 1992 by Peter Stafford
ISBN Pbook: 9780914171195
ISBN Ebook: 9781579512064

Published by:
Ronin Publishing, Inc.
PO Box 3436
Oakland, CA 94609
www.roninpub.com

Technical Editor:	Jeremy Bigwood
Derivitive Manuscript:	Beverly Potter
Molecular Diagrams:	Alexander Shulgin
Cover & Interior Design:	Beverly Potter
Cover Painting:	John W. Allen

Library of Congress Card Number:

Distributed to the book trade by INGRAM/Publishers Group West

Material derived from Chapter Four and Chapter Nine of Psychedelics Encyclopedia, Third Expanded Edition

Thanks and Acknowledgements

This book is derived from experiences and
observations of many associates and dear friends.
Extra special thanks to:

Beverly Potter, John W. Allen, Barry Crombe,
Chuck Silva, Richard Evans Schultes, Timothy Leary,
Arthur Brack, R. Gordon Wasson, Albert Hofmann, Jeremy Bigwood, Humphry Osmond, Jonathan Ott, Clark
Heinrich, Sebastian Orfali, Andrew Weil,
Parlay and Tina, Rojelio Alcorcha, Allan Richardson,
Roger Heim, Stanley Krippner, Ralph Metzner,
Frank Barron, Paul Stamets, J.Q. Jacobs,
Michael B. Smith, Rolf von Eckartsberg,
Richard Alpert, Walter Houston Clark,
Michael Kahn, Terence McKenna,
The Church of the Tree of Life,
Alan Watts, Aldous Huxley,
Robert Anton Wilson,
& Ray Rogers.

Books by Peter Stafford

Psychedelics

Psychedelics Encyclopedia

LSD: The Problem-Solving Psychedelic

co-authored with Bonnie Golightly

Psychedelic Baby Reaches Puberty

Heavenly Highs

Magic Mushrooms

Table of Contents

Teonanacatl

1
History

HERE ARE WELL OVER a hundred psychoactive mushroom species. The number of known mushroom species has risen over the years and has expanded substantially as more have been analyzed. Psilocybin and psilocin molecules are the primary psychedelic agents in the psychoactive mushrooms. However, four related molecules may contribute to mental effects.

The term psilocybian mushrooms has been proposed to include all of the species containing psilocybin, and it will be used in that sense here. Quite distinct isoxazolic molecules are present in the Amanita muscaria, or Fly Agaric, and Amanita pantherina, also known as Panther Caps, which are sometimes said to create psychedelic states. Although the histories of psilocybian and of psychedelic Amanita mushrooms are entwined, these categories are quite different chemically, pharmacologically, and in associated shamanic practices.

Fungi

OF THE HALF MILLION PLANT SPECIES classified, about twenty percent fall into the rather mysterious grouping of fungi. Many botanists consider fungi to be outside the usual concepts of plant or animal. Whatever their classification, some of these mushrooms are capable of causing impressive and often enlightening mental effects in humans.

Fungi are distinguished from ordinary plants in two important ways. First, with a few exceptions they lack the green pigment called chlorophyll that enables plants to make use of light in the production of organic substances. Second, fungi employ microscopic spores rather than seeds for reproduction.

This book is concerned only with the rapidly growing, fleshy Basidiomycetes—the fungi popularly known as mushrooms or toadstools. The gilled "fruiting bodies" or carpophores of a mushroom are the sexual, flowering aspect. The larger part of the plant usually lives underground.

Wasson & Wasson

SCORES OF PSILOCYBIAN MUSHROOMS are of special interest, thanks to the investigations of R. Gordon Wasson and Valentina Pavlovna Wasson. This couple was essential to the discovery that the largest natural production of psychedelics occurs in mushrooms. This revelation came about in 1955, somewhat serendipitously as was Albert Hofmann's discovery of LSD. Their discovery—or "rediscovery"—is worth recounting because it so influenced subsequent developments.

Early Pre-Columbian Use

FOR MILLENNIA, PSILOCYBIAN MUSHROOMS we now know were used by native Americans living mainly in Central America, but also as far south as Chile. While these original mushroom users left few records, they did establish a tradition of psychedelic mushroom use.

What we know of the Indian rites came from gringos. Most mushrooms go by botanical nomenclature that ends with the name of a non-native investigator. Only fragmentary evidence exists relating to earlier generations of mushroom worshippers.

from Sahagun's Florentine Codex

from Magliabechiano Codex

Teonanácatl *drawings from the 16ᵗʰ Century*

The first important clues appearred in 16ᵗʰ Century manuscripts written by Spaniards. Friar Bernardino de Sahagún, who spent most his adult life in Mexico, and Dr. Francisco Hernández, the personal physician to the King of Spain, both described mushrooms used as psychoactive agents in tribal rites in post-Conquest Mexico. In 1598 an educated Indian named Tezozómoc wrote about the ingestion of "inebriating mushrooms" by celebrants at the coronation of Montezuma II.

Divine or Devilish?

IN ADDITION TO VERBAL ACCOUNTS, drawings from Catholicized artists entitled "Teonanácatl"—meaning wondrous, or awesome, or divine mushroom—survive from the 16ᵗʰ Century. One Spanish interpretation portrays a bird-like devil dancing on top

of a mushroom. Another depicts another devil-like
being encouraging an Indian to eat mushrooms.

A Náhuatl Indian dictionary prepared in 1571
distinguished a "mushroom of divine inebriation"
from other nanácatls. Another published in 1885
included the names for several inebriating mushrooms. In a translation of the latter, teonanácatl is
described as a "species of little mushrooms of bad
taste, intoxicating, hallucinogenic."

20ᵗʰ Century Use

ASIDE FROM THESE and other Spanish references, no
effort seems to have been expended trying to identify teonanácatl until the 20ᵗʰ Century. A revival
of interest, strangely enough, began as a scholar's
squabble shortly after an authoritative misidentifi-
cation in 1915.

After a search for teonanácatl in specimens of
Mexican mushrooms, respected American botanist Dr. William E. Safford concluded that there
simply were none. He felt that the Spanish chroniclers must have confused them with dried peyote.
In a talk entitled "Identification of teonanácatl of
the Aztecs with the narcotic cactus Lophophora
williamsii and an account of its ceremonial use in
ancient and modern times," Safford declared that
the dried mescal button resembled "a dried mushroom so remarkably that, at first glance, it will even
deceive a mycologist"! He hypothesized that the
Indians may have deliberately misled the Spanish
in order to protect their use of peyote.

Dissenter

THE FEW SCHOLARS WHO HEARD SAFFORD or later read
his report were mainly hearing about psychoactive

mushrooms for the first time—only to be told that
the mushrooms never existed. But there was one
important dissenter—Dr. Plasius Paul (Blas Pablo) Reko, an Austrian physician who had engaged
in extensive botanical collecting as a hobby while
living in Mexico. Reko had become convinced that
teonanácatl referred to mushrooms, not Safford's
hypothesized dried peyote.

In 1919, Reko published a book entitled El Méx-
ico Antiguo, or The Old Mexico, in which he pro-
claimed that people were still using mushrooms in
Mexico for effectos narcóticos. In 1923 Reko wrote
to Dr. J.N. Rose of the Smithsonian Institution: "I
see in your description of Lophophora, that Dr.
Safford believes this plant to be the teonanácatl of
Sahagún, which is surely wrong. It is actually, as
Sahagún states, a fungus which grows on dung-
heaps and which is still used under the same old
name by the Indians of the Sierra Juárez in Oaxaca
in their religious feasts."

Five years later, journalist/novelist Victor A.
Reko, a cousin of Blas Pablo
Reko, wrote the first published objection to Safford's
thesis. In an imaginative,
popular book written in
1936—Magical Poisons:
Intoxicants and Narcotics
of the New World—he de-
clared that Safford's identification "must be contra-
dicted."

Reko wrote, "The
nanacates are poisonous

Drawn by E.W. Smith, courtesy of Richard Evans Schultes.

Plasius Paul (Blas Pablo) Reko

mushrooms which have nothing to do with peyote.
It is known from olden times that their use induces
intoxication, states of ecstasy, and mental aberra-
tions, but, notwithstanding the dangers attendant
upon their use, people everywhere they grow have
taken advantage of their intoxicating properties up
to the present time."

Victor Reko gave the names "Amanita mexicana" and "Amanita muscaria variant mexicana"
for the mushrooms described in his book. This em-
bellishment of his cousin's views was significant in
attracting renewed attention to a mushroom known
as teonanácatl.

Rediscovery

ALTHOUGH REKO WAS CONSIDERED by many to be only
an amateur, and indeed one given to fantastic ideas,
he nonetheless continued to argue steadfastly that
there were Mexican tribes still using mushrooms
for their shamanic ceremonies. In 1936, more than
two decades after Safford closed the case, Reko
heard from Robert J. Weitlaner, an Austrian-born
engineer who had given up that profession to study
Indian ways. He told Reko that the Otomi Indians
of Puebla, just northeast of Oaxaca, and of nearby
regions were using mushrooms as inebriants and
gave Reko samples of the psychoactive mushrooms.

Reko, in turn, forwarded these samples to Dr.
Carl Gustaf Santesson in Stockholm for chemical
analysis and to the Farlow Herbarium at Harvard
University for botanical examination. Reko's mail-
ing arrived at Harvard in such rotted condition that
the mushrooms were identified only as to the genus
Panaeolus—and perhaps incorrectly so.

The Harvard recipient was the young ethno-botanist Richard Evans Schultes, who had been a medical student until he happened upon Heinrich Klüver's monography on "mescal visions." As Schultes later wrote to Klüver, reading that essay altered his life's course. Schultes changed his doctoral thesis to peyote use on the Kiowa reservation in Oklahoma and thereby began a lifelong interest in mind-changing plants of the New World.

The appearance of Reko's mushrooms out of the blue encouraged Schultes to suggest that these—or something similar—may have been the mushrooms referred to in the Spanish chronicles as teonanácatl. Soon he and a Yale anthropology student, Weston La Barre, began summarizing the available evidence against Safford's arguments. Schultes disputed Safford's conclusion in 1937's Harvard Botanical Museum Leaflets and urged that attention be redirected to identification of the mushrooms.

Mushroom Rites

SCHULTES BEGAN STUDIES WITH REKO in 1938 in northeastern Oaxaca among the Mazatec Indians. They heard reports about the existence of mushroom rites in and near the Oaxacan town of Huautla de Jiménez. They collected specimens of Panaeolus sphinctrinus, which was alleged to be the mushroom

R. Gordon Wasson

John W. Allen

chiefly used in the rites. They also collected specimens of Stropharia—or Psilocybe—cubensis, a mushroom of lesser importance according to the native Mazatecs. These specimens remained in the herbarium at Harvard.

Soon after, Robert Weitlaner's daughter Irmgard and her husband, J.B. Johnson, along with others, attended a midnight mushroom ceremony—or velada—in which the shaman alone was said to have ingested teonanácatl. This ceremony was written up by Johnson for a Swedish journal, and soon forgotten.

Early investigations ended with World War II. Reko went on to other pursuits, and Schultes was sent off to the Amazon to search out rubber sources. Santesson died shortly after completing his chemical analysis, and Johnson was killed in a minefield in North Africa.

Wassons' Contributions

from *Maria Sabina and Her Mazatec Mushroom Velada*/Hans Namuth

Valentina Pavlovna Wasson

THE WASSONS' CONTRIBUTIONS were prompted by an apparently minor incident. R. Gordon Wasson was the American son of an Episcopalian minister who during Prohibition had written Religion and Drink, a book that examined biblical references to the drinking of alcohol by religious figures. In it he took the tact of a fundamentalist, which he was not, and implied that it would be quite unchristian to be critical of alcohol. The royalties of this book enabled the younger Wasson to study in Spain, where he worked as an English teacher—and then for a decade as a financial reporter for the Herald-Tribune. In 1926, Wasson married a Muscovite pediatrician, Valentina Pavlovna.

Devotion Reborn

A YEAR LATER, THE TWO WERE WALKING in the Catskill Mountains when Valentina dashed off the path to gather wild mushrooms. R. Gordon's initial reaction was one of disgust and fear that she might poison herself. He wouldn't even touch these "delicacies." Each of them was surprised to discover that the other had such intense, opposing feelings on the subject.

This incident triggered a lifelong search of cultures manifesting either a great loathing of mushrooms or else a proclivity to treasure them. Soon they were dividing peoples into mycophiles, or mushroom lovers, and mycophobes, or mushroom fearers. Their devotion led them on a worldwide search for references and practices involving mushrooms—in museums, proverbs, myths, legends, folk tales, epics, history, poetry, novels, records made by explorers, and so on.

Twenty-five years later the poet Robert Graves and simultaneously an Italian printer of fine

Mushroom Stone

books, Giovanni Mardersteig of Verona, wrote to
the Wassons about their search. They called attention to the Spanish chroniclers who had touched
on teonanácatl, and referred the Wassons to the
mushroom stones then being discovered in quanti-
ty in the Guatemalan highlands, in El Salvador, and
in southeastern Mexico. Mardersteig sent along a
sketch that he had made at the Rietberg Museum in
Zurich of a mushroom stone, which had first been
published as a photograph in 1908. The meaning of
this object by an unknown artist is uncertain, but it
may speak for generations of mushroom worshippers. More than 300 similar sculptures have since
been found.

A few months later, Eunice V. Pike, a Protestant
missionary to the Mazatecs, informed the Wassons
that the local word for mushrooms meant "the dear
little ones that leap forth." The correspondence
led the Wassons to the leaflets by Schultes, written
some fifteen years earlier.

6,000 Years

WASSON WAS NEVER SURE whether it was he or his

wife who first to
put into words a
hypothesis they
came to agree on
sometime in the
1940s—that as
far back as 6,000
years ago, there
were cultures
that worshipped
mushrooms. The
discovery by then

Two views of a mushroom stone.

Hans Namuth

of the twenty or so mushroom stones seemed to be confirming evidence that the mushroom was the symbol of a religion, like the cross promulgated by Christians, the crescent moon by Muslims, and the Star of David by Jews. Although anthropologists and other experts referred to these artifacts as mushroom stones, they seem to have thought of the term as merely descriptive, regarding these stone carvings as mainly phallic.

Shown opposite here are two views of a mushroom stone in the Namuth collection of the pre-classic Mesoamerican period (1,000-500 B.C.). The figure emerging from the stipe is conjectured to be that of a young woman over a grinding stone or metate.

Expeditions

SCHULTES' PAPERS, which pinpointed the town of Huautla de Jiménez in Oaxaca, gave the Wassons their most important clue about where to look for remaining mushroom cults. They contacted Schultes, who told them about Weitlaner, the Rekos, and the mushroom velada witnessed by Weitlaner's daughter. He even arranged for a guide who had lived with the Indians of Oaxaca. Thus began the Wassons' eight expeditions into the mountains of central Mexico. In their fifties, they undertook these trips in the spirit of pilgrimage.

2
Search for the Sacred

VER THREE SUMMER VACATIONS, the Wassons searched the Oaxacan highlands for someone who could tell them about sacred mushrooms. They spoke to herb venders and collected many species of mushrooms previously unknown to scientists. In retrospect, it is amazing how long it took to find the objects of their search. The problem was that while they did find psychedelic mushrooms, none of which were tried, they found no one who would perform the ceremony or talk about the use of mushrooms. There was no way to tell if they were psychoactive unless tried. Who were they to trust?

In the tiny village of Huautla de Jiménez, where he had traveled ahead of his wife, Wasson came upon the answer when he spoke briefly to Cayetano Garcia Mendoza, a thirty-five-year-old official presiding at the town hall. The date was June 29, 1955. Feeling that he wouldn't have the chance to talk for very long, Wasson asked rather quickly, almost as if in an aside, "Will you help me learn the secrets of the divine mushroom?" Uttering the proper glottal stop at the beginning, Wasson used the term 'nti si tho for the object of his search. The first syllable shows reverence and endearment, the

Allan Richardson

Maria Sabína (right) and her daughter

second expresses "that which springs forth." To his utter surprise the answer this time was: "Nothing could be easier."

Curandera Performs

MENDOZA WAS AS GOOD AS HIS WORD. Later that afternoon he took Wasson to his house where they gathered some of the mushrooms. By evening Mendoza had spoken to the famous curandera Maria Sabína, telling her without further explanation that she should serve Wasson. Wasson then went with New York fashion photographer friend Allan Richardson to a mushroom ceremony.

Richardson had promised his wife that if such an eventuality arose, he wouldn't try the mushrooms. At about 10:30 that evening, he and Wasson

MARÍA SABINA
Her Life and Chants

written by Álvaro Estrada
translation and commentaries by Henry Munn
with a retrospective essay by R. Gordon Wasson
preface by Jerome Rothenberg

Maria Sabína's book.

were each offered a dozen large, acrid species known as Psilocybe caerulescens, or "Landslide" mushrooms, which they consumed over the next hour.

Patterns

"ALLAN AND I WERE DETERMINED TO RESIST ANY EFFECTS they might have," Wasson wrote later, "to observe better the events of the night." In spite of his resolve, he soon began to notice harmonious, geometric patterns that emerged in the dark. Then came visions of palaces and gardens.

THE WONDROUS MUSHROOM
Mycolatry in Mesoamerica

R. GORDON WASSON

Wasson's book

Wasson likened his experience to what the Greeks meant when they created the word ekstasis—a flight of the soul from the body. The experience continued until the very early morning. Wasson and Richardson were the first whites in recorded history to partake of the Mexican divine mushrooms. Wasson described this velada, or night ceremony, fully in his book The Wondrous Mushroom: Mycolatry in Mesoamerica.

Three days after he ingested the sacred mushrooms, Wasson repeated the experience a second time. A few days later, Valentina and their thirteen-year-old daughter Masha tried the mushrooms. Six months later, after returning to New York, Wasson ingested dried specimens and found the effects even more fantastic.

CIA Involvement

MEANWHILE, THE CIA HAD INITIATED A SEARCH for the so-called "stupid bush" and other botanicals that might derange the human mind. The CIA became especially interested in a shrub called piule, whose seeds, they were told, had long been used as inebriants in Mexican religious ceremonies. In early 1953, a scientist from Project ARTICHOKE went to Mexico in search of this plant. Before he left Mexico with bags of plant material, including ten pounds of pi-

ule, he heard wondrous stories about special mushrooms used in connection with religious festivals.

The samples went to chemical labs and the CIA
scientists were excited by the findings. They soon
came upon the Spanish records relating to teonanácatl. Morse Allen, head of the ARTICHOKE
program, was particularly fascinated by indications
that mushrooms could be used "to produce con-
fessions or to locate stolen objects or to predict the
future." Putting high priority on finding the mushrooms, Allen traveled to the
best mushroom-growing
area of Pennsylvania to
enlist potential growers.

Von María Sabina
und dem traditionellen Schamanentum
zur weltweiten Pilzkultur

*German translation of
Maria Sabína's book.*

Shortly after Maria Sabi-
na's 1955 velada, a botanist
informant in Mexico City
sent the CIA a description
of Wasson's discovery. The
report was brief, mainly
indicating that the banker
had envisioned "a multi-
tude of architectural forms"
after he had ingested the mushrooms. That was
enough for the CIA to be interested in the Wassons.

Mobiliztion

THE WASSONS' NEXT EXPEDITION, which took place
during the summer of 1956, was timed for the rainy
season in Oaxaca so they could gather mushrooms.
They were accompanied this time by the French
mycologist Roger Heim, Director of the National
Museum of Natural History in Paris and an ac-
claimed expert on tropical mushroom species. His
role was to supervise the collecting of these fungi

and to determine their taxonomy. Another French-
man, a botanist colleague from the Sorbonne, also
traveled to Oaxaca. Finally, to round out this in-
terdisciplinary team, there was Dr. James Moore, a
chemist from the University of Delaware.

Moore was much more than a mere organic chemist at a university. Known as the CIA's
"short-order cook," he was an expert at synthesiz-
ing psychoactive and chemical weapons for the CIA
on short notice. Moore invited himself along on the
Wassons' return expedition to Mexico. As an entice-
ment, he offered a cash grant from one of the CIA's
cover organizations, the Geschickter Foundation.

Moore did not enjoy the journey or the mushroom ceremony. He said, "I had a terrible cold, we
damned near starved to death, and I itched all over.
There was all this chanting in the dialect. Then they
passed the mushrooms around, and we chewed
them up. I did feel the hallucinogenic effect, although 'disoriented' would be a better word to
describe my reaction."

After the collecting of specimens was complet-
ed, Moore returned to Delaware with a bag of the
sacred mushrooms for analysis. He hoped to isolate
and then synthesize the active principle in large
quantities for the CIA. Sidney Gottlieb, head of
the CIA's chemical wing, noted that if Moore were
successful, it was quite possible that the potentiat-
ing molecules "might remain an Agency secret"—
meaning the analysis wouldn't be published in the
scientific literature, unlike academic discoveries.

Parisian Success

WHILE MOORE WORKED ON THE PROBLEM of extraction
and synthesis, the Heim team succeeded in the

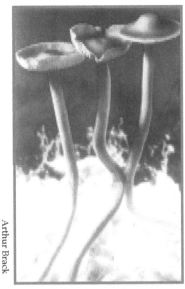

Arthur Brack

Psilocybe mexicana

difficult task of cultivating the species from specimens and spore prints collected in Mexico. Heim asked Sandoz Pharmaceuticals if they would assist in analyzing the mushrooms they grew because his colleagues had been unable to extract the active ingredients. Heim thought Sandoz, successful with LSD-25, might be in the best position to undertake such work. Hofmann accepted Heim's offer with enthusiasm, having already read about the Wassons' discovery.

Heim sent Sandoz 100 grams of dried Psilocybe mexicana, which he had grown in cultures. Trying to establish what would be a reliable dose, the research team tested the mushrooms on dogs. The results were uncertain and almost depleted the mushroom supply. Hofmann then ingested 2.4 grams himself to see if cultivation had ruined its psychoactivity.

Human Testing

WHILE THE DOSAGE WAS MODERATE by Indian standards, the effects led Hofmann to conclude that humans provide a more sensitive testing of mind-affecting substances than do animals. Using about a third of Hofmann's dose, his team members tested various fractionated extracts and isolated 4-OPO_3-DMT and 4-OH-DMT as psychoactive constituents.

Uniquely Qualified

HOFMANN WAS PERHAPS THE SCIENTIST who was best
equipped to analyze the psychedelic agents because
of the considerable chemical similarity between
these substances and LSD. Both contain the same
kind of nucleus with a substitution at the fourth position in the indole ring. "Probably in no other laboratory in the world," wrote Hofmann later, "would
there have been 4-hydroxy indole for comparison
purposes."

He and his colleagues found by extracting the
alkaloids in the mushroom material and degrading
them what the psychoactive principles looked like
chemically. Then they were able to use a route for
synthesis somewhat like that for making LSD. They
published their methods for extraction and synthesis in a Swiss chemical journal, where Hofmann
gave the generic name for the two activating molecules as psilocybin and psilocin, derived from
the Psilocybe genus to which Psilocybe mexicana
belongs.

CIA Fails

THE RAPID ACHIEVEMENT OF HOFMANN'S TEAM ended the
CIA's dream of having its own clandestine reserve.
Though Moore had the mushroom first, he failed
largely because he did not ingest the mushrooms
and therefore could not determine if his mushrooms were still active. Even if he extracted the
constituents in several solutions, he had no way of
telling which solvent contained the active ingredients. There was no reason for him to test for indole
tryptamines. Moore gave up his effort and request-
ed a supply from Sandoz.

Magic Mushrooms Revealed

THE FIRST EXTENSIVE ACCOUNTS of the Wassons' discovery appeared in May 1957 when they published Mushrooms, Russia and History, a monumental two-volume work on their investigations. A shorter, more accessible report of seventeen pages was then published in Life magazine. Only 512 copies of their magnum opus were issued, but the Life article was read by millions.

The third in a "Great Adventures" series, their

Psilocybe mexicana mushrooms

article was titled "Seeking the Magic Mushrooms: a New York banker goes to Mexico's mountains to participate in the age-old ritual of Indians who chew strange growths for visions." The term magic mushrooms was invented by a Life editor. Wasson did not like the name and had reservations about it.

John Marks characterized the tone of the presentation as giving these newly revealed mushrooms "glowing but dignified respect." The article came soon after Huxley's writings about his ingestion of mescaline sulfate and caused quite a stir. Millions were introduced to the mysteries of psychedelics.

Priority Rift

AMONG THOSE MOST INTERESTED in the Life article was Dr. Rolf Singer, a mycologist who carried its illustrations of the sacred mushrooms with him on a two-week trip to Mexico. He was accompanied by two Mexican graduate students in mycology, one being Dr. Gaston Guzmán, who was in the employ of Parke-Davis. This group followed the Wasson/Heim trail, eventually meeting up with Wasson in the remote mountains of Oaxaca.

The Singer group was successful in mushroom collecting. Singer hurriedly published a paper and established scientific priority in the identification of mushrooms that the Heim team had been collecting for years. This haste divided the communities of mycologists and ethnobotanists, creating a rift that continued until the end of the 20th Century.

For all that, the Wassons were still the primary investigators of psilocybian mushrooms. They mobilized an impressive array of interested parties and resources for further study, involving institutions ico. They recruited linguists, chemists, botanists, and other specialists in Mexico, the U.S., Japan, and the Urals— anywhere they found evidence of mushrooms being used as entheogens. In Greek the word entheos means "god within," and gen, "which denotes the action of 'becoming.'"

By the time of Valentina Pavlovna's death from cancer in 1958, most of the teonanácatl story had been uncovered. Commenting later on their search, Wasson said that even if they had been on the wrong track, theirs "must have been a singular false hypothesis to have produced the results that it has."

3
First Wave

LBERT HOFMANN'S WORK led to patents on psilo-cybin and psilocin for Sandoz Pharmaceuticals. After extraction, identification, and synthesis of these naturally occurring molecules, Hofmann and his co-workers synthesized a series of analogs, or related compounds.

These were essentially the same molecules except that the phosphoryl or hydroxy group at the top of the indole ring was moved around to other ring positions, and different numbers of methyl groups (CH$_3$) and other carbon chains were added to the side chains and to the nitrogen on the indole ring to see how these changes would affect psychoactivity. The question of safety was significant because these synthetic substances were being tried for the first time. The new compounds were tested on animals.

Therapy

TWO OF THESE SYNTHETIC COMPOUNDS WERE TESTED in controlled trials in humans and were eventually used in "psycholytic therapy" in several clinics in Europe. CY-19 (4-phosphoryloxy-N, N-diethyltryptamine) and CZ-74 (4-hydroxy-N, N-diethyltryptamine) are the diethyl analogs of psilocybin and psilocin that produce experiences similar to their counterparts, but are slightly less active by weight. Their effects last only about three and a half hours—compared with about twice that for psilocybin and three times that for LSD. These synthetic compounds have been considered particularly appropriate for psychother- apy because they are less tiring, more manageable for the experiencer and therapist, and easier to schedule.

Meanwhile, psilocybin was sent to interested researchers as another psychedelic agent backed by a pharmaceutical house—after Cannabis tinctures, then mescaline, then harmaline and harmine, then LSD-25. Results from psilocybin studies began appearing in 1958, when the conclusions from six research projects were published. By 1962, Wasson listed 362 items of relevant literature in a bibliog- raphy about psilocybin and the mushrooms con- taining this compound. The excitement in experi-

mental psychiatry was intense. News of this psychedelic spread rapidly in the small but expanding drug subculture.

Mushroom Pill

HOFMANN WENT ALONG ON THE 1962 EXPEDITION organized by Wasson to see Maria Sabina. He brought a bottle of psilocybin pills, which Sandoz was marketing under the brand name "Indocybin"—indo

Ampules of early Sandoz psilocybin and psilocin.

Jeremy Bigwood

for both "Indian" and "indole," the nucleus of their chemical structures, and cybin for the main molecular constituent, psilocybin. Psilo in Greek means "bald"; cybe means "head."

Hofmann gave his synthesized teonanácatl to the curandera who divulged the Indians' secret. "Of course," Wasson recalls of the encounter, "Albert Hofmann is so conservative he always gives too little a dose, and it didn't have any effect." Hofmann had a different interpretation. "The pills, which must dissolve in the stomach before they can be absorbed, takes place only after 30 to 45 minutes, in contrast to the mushrooms which, when chewed, work faster because part of the drug is absorbed immediately by the mucosa in the mouth."

In order to settle her doubts about the pills, more were distributed, bringing the total for Maria Sabina, her daughter, and the shaman Don Aurelio up to 30 mg., which was a moderately high dose but not perhaps by Indian standards. At dawn, their Mazatec interpreter reported that Maria Sabina felt there was little difference between the pills and the mushrooms. She thanked Hofmann for the bottle of pills, saying that she would now "be able to serve people even when no mushrooms were available."

Harvard Project

SURPRISINGLY LITTLE OF THE PSILOCYBIN EXPERIMENTATION involved human use. The best known investigations in this area were conducted by Dr. Timothy Leary and his associates at Harvard University—in the same building used by William James when he studied religious mysticism, "laughing gas," or nitrous oxide (N_2O), and the nature of altered consciousness.

Leary, who was Irish and Catholic in extraction and whose father was Eisenhower's dentist, had been rebellious and was expelled from West Point. While Leary was working as research director for Kaiser Hospital in Oakland, he was offered a lectureship at Harvard.

Leary's First Trip

IN THE SUMMER OF 1960, while on vacation in Cuernavaca near Mexico City, Leary ate seven small mushrooms beside the pool of his rented villa. He said he soon felt himself "being swept over the edge of a sensory Niagra into a maelstrom of trans-cendental visions and hallucinations. The next five hours could be described in many extravagant metaphors, but it was above all and without question the deepest religious experience of my life."

Cuernavaca

I was first drugged out of my mind in Cuernavaca, August 1960. I ate seven of the Sacred Mushrooms of Mexico and discovered that beauty, revelation, sensuality, the cellular history of the past, God, the Devil—all lie inside my body, outside my mind.

In the days of Montezuma this town called horn-of-the-cow was the center of soothsayers, wise-men and migicians. Cuernavaca is the southern anchor point of a line running from the fabled volcano of Toluca. On the high slopes of the valcanoes, east and west of the capital, grow the Sacred Mushrooms of Mexico, divinatory fungi, Teonanacatl, flesh of the Gods.

In the summer of 1960 Cuernavaca was the site
of considerable activity by American psychologists from the North vacationing on research
grants and working in the lush valley of More-
los.

Erich Fromm was running an experimental
project down the highway, studying the social
and emotional currents of Indian village life.

Over in Tepoztlan, ten miles to the east, Professor David McClelland, on vacation from
Harvard, was working on plans to help un-
derdeveloped countries raise their economic
standards through psychological techniques
and the Protestant ethic.

Elliot Danzig, Mexico's leading industrial
psychologist, was a few cornfields away in hs
villa which sits next to the cliff under the altar
of the God Tepozteco. It was at this altar, often
shrouded in rain clouds, that the Aztecs has
worshipped the God Tepozteco, to drumrolls of
the arousing thunder and blots of lightning, the
clinging flame. It was he who showered down
blessings including the gift of "pulgue"—a
mikly beer fermented from cactus, which
contains its own abundance of thunder of the
earth.

In Cuneravaca another villa served as summer
headquarters for four American psychologists—Timothy Leary and Richard Alpert of
Harvard, Frank Barron of California, and Richard Dettering of San Francisco.

Many of the scientists who were working and
vacationing there that season had their lives
dramatically changed, and none of them ever
completely escaped from the mysterious power, the challenge, the paradox of what unfolded.

—Timothy Leary, *High Priest*

Leary vowed "to dedicate the rest of my life as a psychologist to the systematic exploration of this new instrument." That fall he returned to Harvard and interested many graduate students and others in researching psilocybin.

Aldous Huxley, who was in Cambridge at the time as a visiting lecturer at M.I.T., was brought in as an advisor. On the day John Kennedy was elected to the Presidency, Huxley and Humphry Osmond visited Leary. They both agreed that Harvard would be a perfect place to conduct a study of psilocybin, but they felt that Leary "might be a bit too square," in Osmond's words.

Leary Contacts Sandoz

IN LATE 1960, LEARY ESTABLISHED an eight-member board to oversee the Harvard Psilocybin Research Project and contacted Sandoz for psilocybin to be used for "creativity studies." The board included Huxley; psychiatrist John Spiegel, who went on to become president of the American Psychiatric Association; David McClelland, the chairman of the social relations department; Frank Barron, an associate who had written much about creativity; Ralph Metzner, who became a close Leary colleague; two graduate students who had already started a project with mescaline; and Leary.

That winter another Harvard faculty member, Richard Alpert, became an important companion. Leary and Alpert were co-lecturers in a course on game theory called "Existential Transactional Behavior Change." McClelland asked Alpert to keep an eye on Leary and his "mushroom project."

Alpert Turns On

IN THE SPRING, ALPERT TOOK SYNTHESIZED PSILOCYBIN—which Leary and others made a point of referring to as "the mushrooms," or "the mushroom pills." His experience turned his life around. According to Alpert, he began by closing his eyes and relaxing. "In the living room of Leary's house in suburban Boston, Alpert saw a figure in academic robes standing a few feet away and recognized himself in his role as Harvard professor. The figure kept changing to other aspects of his identity—musician, pilot, lover, bon vivant—that had somehow dissociated themselves from his body. And then to his horror he watched his body itself disappear as he looked down on it—first his forelegs, then all his limbs, then his torso—and he knew for the first time that there was 'a place where "I" existed independent of social and physical identity… beyond Life and Death.' About five in the morning he walked the few blocks to his parents' house in a driving snowstorm and began shoveling the driveway, laughing aloud with joy…."

Prison Project

THREE PSILOCYBIN PROJECTS WERE LAUNCHED with an emphasis on the psychology of "game-playing." After initial psilocybin investigations, the Leary group began working in early 1961 in nearby Concord with convicts in the Massachusetts Correctional Institution, a maximum-security prison for young offenders.

It was hoped that psilocybin would help prisoners see through the self-defeating "cops-and-robbers game" to become less destructive citizens.

Leary got along well with the Irish warden, and
soon six prisoners volunteered for the study.

> The first psychedelic session at the prison set
> up powerful repercussions. First there was the
> effect on the little group of voyagers. Strong
> bonds had developed. We had been through
> the ordeal together. We had gone beyond the
> games of Harvard psychologist and convict.
> We had stripped off social facade and faced fear
> together and we had trusted and laughed.
>
> —Timothy Leary, *High Priest*

The six volunteers grew in number to thirty-five
over the next two years. Each underwent two psilocybin experiences during six weeks of biweekly
meetings. Leary reported that the prison subjects
were able to detach themselves from their everyday
roles to confront themselves and recognize con-
structive alternatives to their formerly violent and
self-destructive behavior patterns. The question
was what would happen to these prisoners upon
release. Would the insights gained from two fairly
heavy doses of psilocybin help them to lead useful
and rewarding lives? Or would they return to a life
of crime?

Dr. Stanley Krippner, who experimented with
psilocybin at Harvard and who has since worked
in the fields of dream studies and parapsychology,
summed up the results: "Records at Concord State
Prison suggested that 64 percent of the 32 subjects
would return to prison within six months after parole. However, after six months, 25 percent of those
on parole had returned, six for technical parole vio-
lations and two for new offenses. These results are
all the more dramatic when the correctional liter-

ature is surveyed; few short-term projects with prisoners have been effective to even a minor degree.
In addition, the personality test scores indicated a
measurable positive change when pre-psilocybin
and post-psilocybin results were compared."

Intellectuals Project

ALTHOUGH THIS PSILOCYBIN EXPERIMENT included no
control subjects and a lot of "tender, loving care,"
it did establish a sound basis for hope. The results
warranted at least one controlled study. In a second
area of experimentation, Leary and his associates
gave the mushroom pills to about 400 graduate
students, psychologists, religious figures, mathe-
maticians, chemists, writers, artists, musicians, and
other creative individuals to study their reactions.
Extensive records were kept but only a few of these
accounts have been published.

One recording of a Huxley trip mentions only
that he was given 10 mg. of psilocybin and that
he "sat in contemplative calm throughout; occasionally produced relevant epigrams; reported the
experience as an edifying philosophic experience."
Alan Watts' description of the psilocybin experience as "profoundly healing and illuminating" for
him appears in Metzner's The Ecstatic Adventure.
Krippner's account is in Aaronson and Osmond's
book Psychedelics. Barron devotes a chapter in Creativity and Psychological Health to excerpts from
the records made by artists and musicians given
psilocybin.

The Journal of Nervous and Mental Diseases
summarized the findings on the first 129 men and
48 women tested—70 percent considered the experience either pleasant or ecstatic, 88 percent felt

they had learned something from it or had some important insights, 62 percent believed the experience to have changed their life for the better, and 90 percent expressed a desire to try the drug again.

Collaborate with the Patient

After fifteen years of practicing psychotherapy and about ten years of doing research on psychotherapy, I had come to the conclusion that there was very little that one person called a doctor could do for another person called a patient by talking to him across a desk, of listening to him as he lay on a couch.

I was convinced that the doctor had to suspend his role and status as a doctor, had to join the other person actively and collaboratively in figuring out the solution to his problem. As mush as possible, the doctor had to turn over the responsibility to the man who knew most about the problem at hand, namely, the patient.

—Timothy Leary

Mystical Examined

THE RESEMBLANCE OF MYSTICAL EXPERIENCE induced by psilocybin to mystical states brought about by spontaneous rapture or by religious practice is a third area of inquiry which developed out of this work. This eventually became a double-blind study conducted by Walter Pahnke as part of his Ph.D. dissertation for the Harvard Divinity School. Leary described it as a "tested, controlled, scientifically up-to-date kosher experiment on the production of the objectively defined, bona-fide mystic experience as described by Christian visionaries and to be brought about by our ministrations."

Pahnke focused on nine traits listed by Dr. W.T. Stace, Professor Emeritus at Princeton, which Stace felt were the fundamentals of mystical experience—"universal and not restricted to any particular religion or culture."

In a Boston University chapel on Good Friday, twenty Christian theology students took part in Pahnke's experiment in 1962 after having been exhaustively tested and screened. Ten were given 30 mg. of psilocybin. The others received 200 mg. of nicotinic acid and a small amount of benzedrine to stimulate the initial physical sensations of a psychedelic. Neither the subjects nor their guides knew—at first—which drugs had been given to whom.

Miracle of Marsh Chapel

THE EXPERIMENT BECAME KNOWN as "The Miracle of Marsh Chapel." In the following six months, extensive data were collected, including tape recordings, group discussions, follow-up interviews, and a 147-item questionnaire used to quantify characteristics of mystical phenomena.

The results revealed that the reaction level in each of Stace's categories was significantly higher for the psilocybin group than for the controls. Nine out of the ten who ingested psilocybin reported having religious experiences that they considered authentic. Only one control claimed to have had even minimal spiritual cognition. More important in terms of genuine mystical experience, there was a lasting effect upon behavior and attitudes.

Pahnke summarized these results: "After an admittedly short follow-up period of only six months, life-enhancing and life-enriching effects, similar to some of those claimed by mystics, were shown by the

Stace's Nine Fundamentals of Mystical Experience

Experience of unity
Experience of timelessness
and spacelessness
Sense of having encountered
ultimate reality
Feeling of blessedness and
peace
Sense of the holy and the divine
Experience of paradoxicality
Sense of ineffability
Transciency
Persisting positive changes in
attitude and/or behavior

higher scores of the experimental subjects when compared to the controls. In addition, after four hours of follow-up interviews with each subject, the experimenter was left with the impression that the experience had made a profound impact, especially in terms of religious feeling and thinking, on the lives of eight out of ten of the subjects who had been given psilocybin…. The direction of change was toward more integrated, self-actualized attitudes, and behavior in life."

Subsequent requests by Pahnke for psilocybin and government approval to repeat his study were denied.

4
Grow Your Own

SYNTHETIC PSILOCYBIN WAS AVAILABLE to only a small number of those in the drug subculture and some with academic connections during the 1960s because it was expensive and complicated to make. As for the organic product, it was widely believed then that sacred mushrooms grew only in Mexico. Most who experienced them had to travel to Huautla de Jiménez to do so. The Turn-On Book and The Psychedelic Guide to the Preparation of the Eucharist described methods for growing Psilocybe cubensis on agar and grains, but these techniques went largely untried because they were not very effective and did not produce many mushrooms.

Pop Culture Fueled

As ADDITIONAL SPECIES OF MUSHROOMS containing psilocybin and psilocin were discovered, biosynthetic studies utilizing Psilocybe cubensis were initiated in many university laboratories. Mycologists as a group were uninterested in publicizing psychoactivity.

Only a few people having access to the mycological literature knew that Psilocybe semilanceata, which Heim and Hofmann had analyzed as being quite potent, grew extensively along the coast of the Pacific Northwest. The main clues appeared in French, in Heim and Wasson's Les Champignons Hallucinogenes du Mexique and Nouvelles Investigations sur les Champignons Hallucinogenes.

With the increasing anti-LSD propaganda, interest in organic or "natural" psychedelics greatly increased. The appeal was fueled by Leary's book High Priest, which devoted attention to mushrooms and psilocybin, and by the wildly popular books of Carlos Castaneda. The first of these, The Teachings of Don Juan: A Yacqui Way of Knowledge, caught the popular imagination as many readers became fascinated with Don Juan's "smoking mixture"—allegedly a combination of mushrooms, Datura, and other substances sealed in a gourd for more than a year.

Mushrooms In Demand

IN ORDER TO MEET THE NEW DEMAND for magic mushrooms, many dealers simply renamed their LSD, PCP, or other compounds, some even claiming that

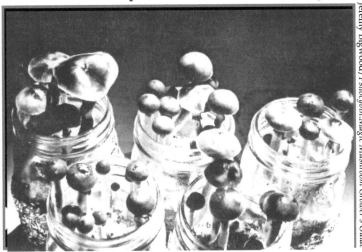

Jeremy Bigwood / Psilocybin:Magic Mushroom Grower's Guide

Jars of Stropharia cubensis

their psilocybin had been cut with "organic rose hips," apparently to make it sound even more "natural." Unscrupulous dealers adulterated Agaricus bisporus—the common, edible, commercial mushroom, which is now called A. brunnescens—with LSD or PCP or both, a practice that continued for several years.

PharmChem, after analyzing hundreds of suspect mushroom samples, reported that it found only two genuine specimens over a three-year period. Bruce Radcliff of PharmChem dubbed the others Pseudopsilocybe hofmannii.

Hunt Your Own

A NEW ERA IN HUNTING MUSHROOMS opened in the 1970s after publication of Leonard Enos' book, A Key to the American Psilocybin Mushroom, which described fifteen species in sixty pages, along with a water-color picture of each. Enos had personal experience with only two of the species and, thus, his renderings of the mushrooms' appearances were inaccurate and sometimes fanciful. In two cases, species known by two Latin binomials were drawn to look like different kinds of mushrooms.

Enos also provided an overcomplicated section on cultivation, which no one seems to have used. Nonetheless, Enos' book stimulated much American fieldwork that resulted in the publication of several reliable guides, such as the Hallucinogenic and Poisonous Mushroom Field Guide by Gary Menser.

Best Opportunity

A GROUP OF A DOZEN PEOPLE in the San Francisco Bay Area began experimented in the late 1970s with various techniques for growing psilocybian mushrooms, hoping to find a simple procedure for cultivation. They had already grown several Banisteriopsis species and the San Pedro cactus in a search for a natural psychedelic that could be grown indoors and offered a high yield. They concluded that mushrooms presented the best opportunity.

Writing under the names O.T. Oss and O.N. Oeric, this group published the results of their mushroom experimentation in Psilocybin: Magic Mushroom Grower's Guide. This book, written in plain language and accompanied by photographs, described a fairly simple and effective technique for home cultivation of the potent Stropharia (Psilocybe) cubensis species that required no controlled materials. Spores were available through counterculture magazines such as High Times.

Oss & Oeric Method

THE IMPORTANT DISTINCTIONS in the Oss and Oeric method is their instructions for maintaining sterile conditions and with "casing." J.P. San Antonio by then had published a new laboratory procedure for producing Agaricus brunnescens in small amounts for scientific study. He showed that covering the mycelium in the vegetative stage with about half an inch of slightly alkaline soil greatly increases the yield by causing it to fruit repeatedly in flushes appearing periodically.

Although it is unclear who deserves the credit for this breakthrough, the San Antonio technique was modified so that it worked with Stropharia cubensis grown on rye and other grains.

Cultivating Psilocybian Mushrooms the Old-Fashioned Way

(1) Equipment need-
ed for scalpel isola-
tion of sterile flesh of
Stropharia cubensis.

(2) Step-by-step
growth from spores
to mushrooms.

(3) Inoculations of
Mason jars filled
with rye and water
from stock culture.

(4) Mycelial growth after five, ten, and fifteen days.

(5) Casing and spraying Mason jars.

(6) Fresh mushrooms and dried mushrooms ready for freezing.

Modern Psilocybin Cultivation

Mason Jars are rarely used any more for cultivation. The following photo set shows the steps in modern psilocybian mushroom

Sterilzed rye berry seeds are used to grow the mycelium from spores dropped by the release of a valve into a bag of such

Photos by John W. Allen

(Left) The mycelium spreads through and feeds off the rye berry seeds in the plastic bag. (Right) When the bag has completely turned into pure white mycelium it is ready to be transferred to a compost mixture.

In seven to thirteen days mushrooms begin to appear.

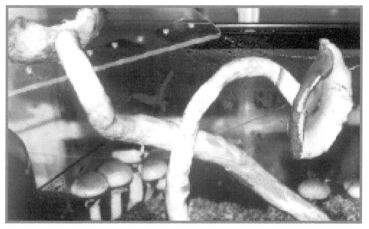

Two 23 inch long Orissa, India Psilocybe cubensis specimens in the aquarium. The smaller "shrooms" in the background are approximately 8 inches i height.

The spawn of mycelia and rye berry are mixed with a five-pound bag of pasturized wheat straw, then added to a terrarium with air holes or an aquarium for propagation.

Oss and Oeric's goal was to enable mushroom lovers to cultivate these mushrooms in their homes, independent of the illicit market. They hoped that this technique would become a permanent part of the subculture, immune from surveillance and anti-drug crusades. To a certain extent, these ends were met.

Complicated

THE OSS AND OERIC TECHNIQUE was so complicated that only a small percentage of people ever used it. A large pressure-cooker was needed. Many couldn't get the hang of the spore-growing and sterilization requirements for inoculation of the rye jars.

Since a capital investment of only about $200 enabled one to produce large volumes, some people took on the procedure as a full-time job, producing thousands of jars of these mushrooms. Some in the Oss and Oeric group were disappointed by this cottage industry because it resulted in a kind of centralized capitalism.

Many cultivation guides are available.

Since publication of the Oss and Oeric book, several others have issued procedures for psilocybian cultivation. Many mushroom-growing kits were presented to the public. "Dung dealers" even offered high-priced compost for sale.

Over this period, many interested parties became knowledgeable

Over this period, many interested parties became knowledgeable about mushroom growing, with a few people doing much additional experimentation. Among other developments, there was a fair amount of cultivation of other psilocybian species such as of Psilocybe cyanescens, which is large and full of psilocybin, but a poor fruiter.

5
Harvesting & Preparation

USHROOMS CONTAINING PSILOCYBIN and/or psilocin belong to a broad botanical order, the Agaricales, or gill-bearing fungi, and are found mainly in the Psilocybe, Stropharia, and Panaeolus genera. Over one hundred species of psilocybian mushrooms are known, each exhibiting distinct ranges in size, shape, habitat, and potency. Potency goes from virtually nothing to approximately 15 mg./gm. of the dry weight. The average is perhaps 3.0 mg./dry gm., amounting to about 0.03 percent of the fresh mushrooms. Mushrooms are generally 90 percent water.

When the Wassons asked their Indian guide about the source of mushrooms, he replied, "The little mushroom comes of itself, no one knows whence, like the wind that comes we know not whence nor why." Actually, psilocybian mushrooms begin as microscopic spores that grow on the tip of cells near the mushrooms gills, called basidia. After maturing, they are dispersed by the wind.

Out of perhaps a million such spores, only a few reach an appropriate habitat and grow. They develop threadlike into hyphae, thin cells that mass together and spread underground to become the

mycelium, which corresponds to the leaves and
roots of a green plant. When this structure fruits,
the sexual part of the organism appears above
ground as a mushroom.

Rain, Wind & Germination

As THE MUSHROOM EXPANDS through the absorption
of water by osmotic action, a protective veil develops under the gills. Eventually the cap breaks the
veil through an evaporation process caused by the
sun, and the veil and stem are covered with spores,
usually a dark purple-brown. Millions of spores
are carried by the rain and wind to where they will
germinate and again produce mycelia, thus completing another cycle.

Not all mushrooms are safe to consume. Anyone interested in collecting or growing psilocybian
mushrooms should, of course, acquire a field guide
by a competent author, such as Paul Stamet's Psilocybe Mushrooms & Their Allies. Additional valuable illustrations, photographs, and descriptions
can be found in Ott and Bigood's contribution to
Teonanácatl, Gary Menser's Hallucinogenic and
Poisonous Mushroom Field Guide, and Richard
and Karen Haard's Poisonous & Hallucinogenic
Mushrooms.

Harvesting

WHEN COLLECTING WILD MUSHROOMS, grasp and twist
only the stem, not the cap. All species should be
separated in wax paper or paper bags. They should
not be placed in plastic baggies because mushrooms must breathe or they rapidly spoil and de-
compose. Care should also be taken to see that they
aren't crushed.

Notes made about the mushrooms' habitat often give valuable clues to identification of psilocybian mushrooms, as does a spore print. This can be made by placing a specimen's cap on a sheet of white paper and, possibly, another on black paper. These mushroom caps are then covered by a glass and left alone for several hours, until it becomes clear whether or not they are dark purple-brown. The remainder of the collection should be refrigerated as soon as possible, but not frozen, because mushrooms deteriorate rapidly when thawed. They can be preserved in the vegetable bin of a refrigerator for about a week.

Bluing

SOME PSILOCYBIAN MUSHROOMS EXHIBIT a striking blue color characteristic in fresh specimens. This can aid in identification along with other traits, such as

Mushrooms bluing

spore color and size, appearance of the gills, and so forth. When these mushrooms are scratched or bruised by handling, they stain blue or, if the surface color is yellowish, they stain greenish blue. Some of these mushrooms exhibit this stain naturally, perhaps because of the heat of the sun or the pressure of raindrops.

There has been much mention of this test in the psychedelic literature, but it is by no means reliable. Some mushrooms "blue" only after the first flush of the fruiting bodies. Growth of the fruiting bodies recurs, usually at one-week intervals. Some authors have concluded that this "bluing" occurs only in mushrooms containing both psilocin and psilocybin.

It is unclear which tryptamine in the psilocybian mushrooms is responsible for the bluing. Pure psilocybin or psilocin when placed in pure water and left at room temperature discolors the water to a bluish-brown. Jeremy Bigwood and Michael Beug, after analyzing many collections of fifteen psilocybian species, concluded that bluing per se does not indicate the presence of psilocin.

A few species containing only psilocybin exhibited a slight bluing reaction on the stems but would not blue when handled. The species containing both the 4-phosophorylated psilocybin and the 4-hydroxylated psilocin analogs, however, bruised a darker blue and often showed this characteristic even when untouched.

Preparation for Storage

HARVESTED PSILOCYBIAN MUSHROOMS CAN BE EATEN FRESH, or they can be dried, sealed, and stored. The best procedure is to dry the mushrooms in a freeze-drier without heat. For most users, this is impossible, so a lamp or oven will do, as long as there is ventilation and the temperature does not exceed 90° F. in a dry atmosphere.

When an oven is used, the door should be left open a crack. A space heater may also be used. Whatever the means, the drying should take at

Psilocybe cubensis mushrooms, dried and bagged.

least twenty to twenty-four hours and leave the mushroom in a brittle state.

The mushrooms may then be weighed, placed in a sealable container, and frozen. Mushrooms are almost never frozen fresh because they then disintegrate when thawed.

Grind & Freeze

To produce a homogenous mixture from which known doses can be accurately weighed, the mushrooms could be ground in a blender or coffee mill. The resulting powder might be best stored frozen at the lowest temperature possible in an airtight container filled to the top. The result doesn't look like mushrooms, and probably cannot be identified by species even by a mycologist. It is quite easy to measure out doses for ingestion.

Freezing is most critical for those ground mushrooms known to contain psilocin, such as Psilocybe

Freezing is most critical for those ground mushrooms known to contain psilocin, such as Psilocybe cyanescens or Stropharia cubensis, because they have a short shelf life at room temperature. For ingestion, such powders could be capped, blended into a "smoothie," or drunk with chocolate. A chocolate drink prepared with honey, spices, and water—there was no milk in pre-Conquest America aside from corn whey—has long been associated with mushroom rituals, and is quite pleasant served before a velada.

6
Chemistry

THE PSYCHOACTIVE COMPOUNDS in psilocybian mushrooms are psilocybin, psilocin, and their N-de and N-di-demethylated analogs. Workers at Sandoz Pharmaceuticals and elsewhere have synthesized many related compounds. But only two have been tested in humans, which were lab-coded "CZ-74" and "CY-19."

The Indole

ALL SUCH COMPOUNDS contain the white, crystalline ring structure of an indole which chemists abbreviate as C_8H_7N. Psilocybian mushrooms also contain ethylamine side chains of various lengths. Taken together, the indole and side chain constitute tryptamines.

Idole

LIKE Brain Chemicals

THE PSILOCYBIN AND PSILOCIN molecular grouping bears a close resemblance to chemicals found in the brain. One of the psilocybian analogs is, in fact, one of the closest known compounds to the neurotransmitter serotonin, differing only with respect to the rare substitution just mentioned at position 7.

4-OPO₃-DMT (Psilocybin)
(4-phosphoryloxy-N, N—
dimethyltryptamine)

4-OH-DMT (Psilocin)
(4-hydroxy-N, N-
dimethyltryptamine)

Psilocybin (left) is the main psychoactive compound in psilocybian mushrooms. Most of it is transformed upon ingestion by the enzyme alkaline phosphatase into psilocybin (right), the second important psychedelic in these mushrooms.

5-OH-T, also called 5HT
(Serotonin)

5-OH DMT (Bufotenine)

Close cousins to psilocin are bufotenine (right), a substance in plants and animals that was once thought psychoactive, and the human neurotransmitter serotonin (left).

4-OPO₃-DET (CY-19)
(4-phosphoryloxy N, N-
diethyltryptamine)

4-OH-DET (CZ-74)
(4-hydroxy-N, N-
diethyltryptamine)

Two synthetics in this clustering that have been tested in people with good results are CY-19 and CZ-74.

Related compounds present in some psilocybian mushrooms are 4-OPO₃-NMT (baeocystin) and 4-OPO₃-T—both analogues of psilocybin—and 4-OH-NMT (norbaeocystin) and 4-OH-T—both analogues of psilocin.

It is 4- rather than 5-hydroxytryptamine. Psilocin, interestingly, is the nearest relative to bufotenine, once thought to be a psychoactive compound, which was first discovered in the skin secretions of toads—Bufo vulgaris, for which it was named— and later in plants, notably the tree known as Anadananthera peregrina from which cohoba snuff is made.

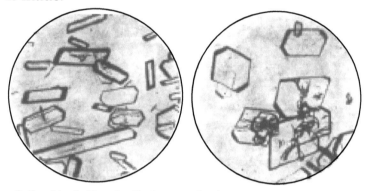

Psilocybin (left) and psilocin crystals viewed by microscope Reprinted *from* Discoveries in Biological Psychiatry.

Psilocybin and Psilocin

THE MAJOR PSYCHEDELIC AGENT in psilocybian mush-rooms is psilocybin—the first indole derivative dis-covered to contain phosphorus. When ingested the phosphorus radical is immediately "dephosphor-ylated" by an intestinal enzyme, called alkaline phosphatase, into psilocin and phosphoric acid. Animal experiments suggest that psilocybin and psilocin appear at similar chemical concentrations at about the same time in various organs. Thus, the phosphorus radical is generally considered dead weight in terms of psychoactivity.

An important difference between psilocybin and psilocin is their relative stability—psilocin is easily oxidized, deteriorating soon afterwards. For this reason Sandoz chose to develop psilocybin, which doesn't require freezing to retain potency, rather than the easier-to-synthesize psilocin. Unfortunately, mushrooms containing just psilocybin and its analogs—Psilocybe semilanceata, Psilocybe pelliculosa and Psilocybe mexicana for example—are all tiny. By weight, psilocin is about 1.4 times as strong as psilocybin—a ratio corresponding to their molecular weights. Compared by weight, LSD is about 200 times as powerful as psilocybin.

Psychoactive Derivatives

MANY USERS FEEL THAT PSILOCYBIN and psilocin in synthetic form produce a more lucid mental state than the mushrooms. They also seem to provide more physical energy. Mushrooms generally have longer effects and are more sedating. These differences are probably caused by the presence of the psilocybin and psilocin analogs, which appear in small amounts but may act as sedatives. These analogs have been tested only on animals.

7
Mushroom High

HE AMOUNT OF TIME REQUIRED to produce somatic sensations from psilocybin, psilocin,
or psilocybian mushrooms varies with the
mode of ingestion. If a high dose of mushrooms is chewed well and kept for some
time in the mouth, effects may be perceived within seven to eight minutes. Psilocybin or psilocin
placed under the tongue—or moderate amounts of
the mushrooms retained for a while in the mouth—
produces initial sensations within about fifteen to
twenty minutes. If the mushrooms are immediately
swallowed, however, only about half of the poten-
tiating chemicals are absorbed by the stomach wall.
It then takes thirty to forty-five minutes, maybe
up to a full hour, before they cross the blood-brain
barrier to prompt psychoactivity. When psilocybin
and psilocin are injected intramuscularly, the effects
are felt within five to six minutes.

Take-Off

GENERALLY, THE FIRST SIGNS that the effects are starting are involuntary yawning—usually without
sleepiness—and a non-specific sense of restlessness
or malaise. Some people experience nausea with
mushrooms, most often after they ingest the bit-

ter Psilocybe caerulescens, Psilocybe aztecorum, or similar acrid species. A few users feel a chill as the effects come on, weakness in the legs, or slight stomach discomfort. Others feel drowsy and may want to curl up and go to sleep.

In most instances, yawning and a slight sensation of physical disorientation or giddiness are the characteristic experiences during the short take-off stage, which usually lasts half an hour and is followed by feelings of lightness and physical harmony. For some, bodily discomfort persists much longer.

High-dose studies of rats suggest that psilocin taken orally is distributed throughout the body. Concentrations in tissues appear highest about half an hour after ingestion, decreasing rapidly over the next three to four hours. The adrenal glands of the test animals show the highest concentrations after the first hour. Until then, the kidneys have more. The small intestine, skin, bone marrow, lungs, stomach, and salivary glands also have significant concentrations—greater, in fact, than those in the brain.

Physical Reactions

As with LSD, PSILOCYBIN AND PSILOCIN prompt few obvious physiological reactions in most people—dilated pupils and, in some users, a sensation of dry mouth. A few studies have noted a slight rise in blood pressure, heart rate, and temperature, but these reactions most often appear to result mainly from apprehensions about the experience or from environmental factors.

Psilocin's primary physiologic effect upon the brain seems to be inhibition of the neurotransmitter serotonin, an effect resembling that brought about by LSD. This finding, together with a notable cross-tolerance exhibited between psilocin and LSD, suggests that both compounds act upon similar mechanisms—or possibly, as Hofmann has put it, on "mechanisms acting through a common final pathway."

Tolerance

SUBSTANTIAL TOLERANCE CAN BE built up by repeated doses taken in close sequence. Dr. Leo Hollister gave psilocybin to a subject on a daily basis for twenty-one days, starting with 1.5 mg. and increasing it to 27 mg. On the twenty-second day, the subject showed hardly any reaction to 15 mg. After a rest of several weeks, however, the same dose produced the normal degree of psychoactivity.

In general use, psilocin tolerance tends to be less pronounced than is the case with LSD. Furthermore, cross-tolerance between these two compounds, when ingested alternately, is not as complete as tolerance developed after repeated ingestion of just one alone.

Duration in Body

IN THE HIGH-DOSE STUDY OF RATS, all but 6 percent of the psilocybin was excreted within twenty-four hours. In humans, only 80 to 85 percent psilocybin and its metabolites is excreted within eight hours: about 65 percent in the urine, and about 15 to 20 percent in bile and feces. Some 15 to 20 percent lingers on, stored in fatty tissues. Significant quantities appear in urine up to a week later. A full 25 percent of the originally administered dose enters urine as psilocin.

Toxicity

PSILOCYBIN, PSILOCIN, AND PSILOCYBIAN mushrooms
show low toxicity. Doses up to 200 mg. of psilocy-
bin/kg. of body weight have been given to mice
intravenously without lethal effects. When dosage
was increased to 250 mg./kg. of body weight, a
few of the mice died. In terms of average human
weight, this corresponds to about 17 or 18 gm. of
psilocybin—more than 2,000 times the dosage recommended by Sandoz when it originally marketed
this drug.

Extrapolation from animal studies to human use
gives only a rough estimate of toxicity. It does seem
safe to say that one would have to consume well
over a kilogram of the most potent fresh Stropharia
cubensis mushrooms—which vomiting would pre-
vent—even to approach the lethal range. Jonathan
Ott, an organizer of mushroom conferences and
author of Hallucinogenic Plants of North America,

*John W. Allen, Jonathan Ott and Paul Stamets, leading
experts in field identification and studies of entheogenic mushrooms.*

said, "I know of no case where an adult has been
made seriously ill by psilocybian mushrooms. Tens
of thousands of intentional inebriations occur each
year with psilocybian mushrooms, in the Pacific
Northwest alone, yet no conspicuous medical prob-
lems has emerged."

Mental Effects

PSILOCYBIAN MUSHROOMS, PSILOCYBIN, AND PSILOCIN can
produce profound, awesome effects upon the mind.
Subjects given psilocybin and other drugs in blind
experiments were generally unable to distinguish
this substance from LSD or mescaline of compara-
ble dosage until several hours had passed. Recog-
nizing psilocybin at that point was possible because
of the shorter duration of its psychedelic effects.

In terms of timing, psilocybian experience is
characterized by the user's going up rapidly, nearly
always achieving an enlargement in the scope of
perception that persists for about two hours. Then
a gradual decline is experienced over the next three
to four hours, resulting in the restoration of ordinary consciousness. The quality of thoughts and
feelings evoked during these two phases varies
greatly, according to mental set, surroundings, and
the dosage ingested.

Oral V. Injection

LEO HOLLISTER AND HIS ASSOCIATES at the Palo Alto
Veterans Hospital reported on blind experiments
in which they gave psilocybin orally and by injec-
tion to a group of "psychologically sophisticated
volunteers," such as graduate students. They found
that the threshold oral dose was about 60 mcg./kg.
of body weight—about 4 gm. for an average-sized

person—"from which minimal but definite changes were produced." An appropriate comparison measure for drugs like amphetamines or barbiturates is perhaps body weight. In contrast, drugs stimulating the central nervous system, such as psilocybin and other psychedelics, probably vary in effect more directly in terms of brain weight.

Opportunity Lost

IN THEIR OBSERVATION OF PSILOCYBIN EXPERIENCES, Hollister and his associates paid almost no attention to such factors as the mental set of individual subjects—despite their background in psychology—or setting, except to say that "the visual beauty of the colored images, especially when augmented by the stroboscopic light during the electroencephalogram, seemed to be a mystical experience to some." The account of mental effects, unfortunately, consists of about 130 words, collected on the next page, from Chemical Psychoses:

Mental Effects

Alterations in mood, either euphoria or dysphoria

Disturbed concentration and attention

Psychological functioning is impaired

Blurred vision, brighter colors, longer afterimages, sharp definition of objects, visual patterns with eyes closed

Increased acuity of hearing

Dreamy state, slow thinking, feelings of unreality, depersonalization

Incoordination, difficult and tremulous speech

Colored patterns and shapes, generally pleasing, sometimes frightening, most often with eyes closed, occasionally superimposed upon objects in field of vision

Undulation or wavelike motion of viewed surfaces

Euphoria, general stimulation, ruminative state

Slowed passage of time

Transient sexual feelings and synesthesias

A few auditory hallucinations

Changes in the body image, the extremities appearing larger than normal

8
The Great Oracle

HE EARLIEST REPORT OF MUSHROOM ingestion comes from Tezozómoc, who commented on the celebrants seeing visions and hearing voices at the coronation of Montezuma II. "Therefore they took these hallucinations as divine notices, revelations of the future, and augury of things to come." Indian users have traditionally employed the visualization and vocalization in psilocybian experiences for purposes of divination, prophecy, healing, and worship.

Mushroom Oracle

Perhaps there is illness in the family and the mushroom is consulted to learn whether the patient will live or die. If the verdict is for death, the family does not wait but immediately prepares for the funeral, and the sick person loses the will to live and shortly afterward gives up the ghost. If the verdict is for life, the mushroom will tell what must be done if the patient is to recover. Or, again, if a donkey has been lost or if some money has been stolen, the mushroom is consulted and gives the answers. Among these unlettered folk, speaking languages that are not written,

there is often no news of an absent member
of the family, perhaps one who has gone as
a 'wetback' to the United States. Here the
mushroom, as a postal service, brings tid-
ings of the absent one, whether he is alive
and well, or sick or in jail, or prosperous
or poor, or whether he is married and has
children.

—R. Gordon Wasson

Psychic Experiences

WASSON HAD SPECIAL REASON TO BE INTERESTED. During
the first ceremony, Maria Sabina asked him what
question he wanted answered. After fumbling
about, he asked about his son in Cambridge. She
had never heard of the place. Later that evening,
she reported that his son wasn't in Cambridge but
at home, that he was in emotional turmoil over
a girlfriend, and was about to join the Army. Although Wasson knew nothing of this at the time, all
turn out later to be true.

Krippner, one of the subjects of the Harvard
Psilocybin Project, gives another example of a psy-
chic-knowing while high on mushrooms: "I found
myself gazing at a statue of Lincoln. The statue was
entirely black, and the head was bowed. There was a
gun at the base of the statue and someone murmured,
'He was shot. The President was shot.' A whisp of
smoke rose into the air. Lincoln's features slowly fad-
ed away, and those of Kennedy took their place. The
setting was still Washington, D.C. The gun was still at
the base of the statue. A wisp of smoke seeped from
the barrel and curled into the air. The voice repeated, 'He was shot. The President was shot.' My eyes
opened; they were filled with tears…."

Of this, he wrote later: "In 1962, when I had my first psilocybin experience, I gave this visualization of Kennedy relatively little thought, as so many other impressions came my way. However, it was the only one of my visualizations that brought tears to my eyes, so I described it fully in the report I sent to Harvard. Nineteen months later, on November 23, 1963, the visualization came back to me as I mourned Kennedy's assassination."

Creativity Stimulus

HOFMANN HAS BEEN QUOTED about the Mexican tone coloring his first mushroom experience—when the Germanic doctor hovering over him appeared as an Aztec priest. In Hofmann's psilocybin experiments, Mrs. Li Gelpke, an artist, also participated. About a drawing she made at that time, she said, "Nothing on this page is consciously fashioned. While I worked on it, the memory [of the experience under

psilocybin] was again reality, and led me at every stroke. For that reason the picture is as many-layered as this memory, and the figure at the lower right is really the captive of its dream…. When books about Mexican art came into my hands three weeks later, I again found the motifs of my visions there with a sudden start."

Similar phenomena have been noted by Wasson, who has conjectured that ancient Mexican art may have been influenced by visionary images appearing during mushroom sessions.

9
Dosage

LONG WITH MENTAL SET AND SETTING, dosage is a major consideration in the quality of a psilocybian experience. Albert Hofmann's view has been that the medium oral dose for psilocybin is 4-8 mg., which "elicits the same symptoms as the consumption of about 2 gm. of dried Psilocybe mexicana fungus." When Sandoz first distributed psilocybin, the pills contained 2 gm. each. They recommended four to five of these in conjunction with "psycholytic" psychotherapy.

Mushroom Pills

AFTER THE MUSHROOM PILLS ARRIVED AT HARVARD, Leary's group quickly discovered that larger amounts produced more impressive results. Leary and a colleague took 20 mg., and a third person consumed 22 mg. In the first session of the Concord prison project, Leary took 14 mg., while the three volunteer convicts took 20 mg. each. By the time Michael Hollingshead arrived with his bottle of LSD, the Harvard group was using as many as three 20 mg. pills for each trip.

Hollingshead therefore took three. He said, "There was a certain amount of intensification of colors, but nothing compared to LSD. So then I took 100. And then, though it was a shorter time, it was very impressive." On the subject of appropriate dosage, there is clearly some distance between Hollingshead and Hofmann.

Regarding native use of psilocybian mushrooms, Wasson reports that usually each adult Indian is given four, five, six, or thirteen pairs. Thirteen pairs is common because thirteen is considered a lucky number. With the wide variations in both mushroom size and potency, such rough guidelines undoubtedly result in enormous differences in the amounts of psilocybin and psilocin consumed.

First-Time Tripper

QUESTIONS AS TO THE PROPER DOSE FOR A FIRST TRIP are difficult to answer, and no answer will be attempted here. With individual variations in mental set and setting taken into account, any general recommendation is bound to be too high or too low in a significant number of cases. It is up to the initiate to decide whether he or she should seek a more than recreational experience the first time out.

Experienced Trippers

FOR EXPERIENCED USERS, Rolf von Eckartsberg illustrates a point about regulating the quality of the experience through dosage in "To Be Able to Say: Thou, Really to Love," reprinted in The Ecstatic Adventure. Von Eckartsberg and his wife took three low-dose psilocybin pills. Soon he became aware that she "was floating through space, giggling, squirming, fluttering like a butterfly." He felt incapable of any emotional reactions. "I found myself standing apart," he reported, "removed by worlds, only half real, half empty and half dead." He pulled himself out of this "peculiar lack of emotional underpinning" by taking "two

Jeremy Bigwood

Psilocybe caerulescens, the sacred mushroom tried by Wasson.

more pills, one after the other, at about forty-minute intervals."

The additional propulsion soon resulted in "a wonderful openness. I am held in the grasp of a comprehensive clarity, lucidity, like very clear, warm, transparent glass." By the end of his report, von Eckartsberg declared, "For the first time I feel like a complete human being, centered in myself, yet an open platform, nothing to hide, completely reconciled and in harmony, a true partner, a steady pole…."

Dosages increased beyond a certain threshold can significantly alter psilocybian experiences, which range from heightened sensitivity to cold and sound through feelings of "mental stillness" and acuity to enhanced rapport with others and mystical states. Amounts of psilocybin above 8 or 10 mg. can produce the same gamut of experience available with LSD. In psychotherapy, doses of 10 mg. of psilocybin and over have been used to good effect in penetrating the defenses of compulsive-obsessive patients, in aiding transference, and in reviving childhood memories for the purpose of dealing with early traumas.

Mysticism

IF MENTAL SET AND SETTING ARE SACRAMENTAL, the results can be mystical, as with Wasson's first mushroom experience. However, circumstances need not be exceptional to evoke impressive responses, as Leary learned after eating seven small mushrooms beside a pool. "The discovery that the human brain possesses an infinity of potentialities and can operate at unexpected space-time dimensions left me feeling exhilarated, awed, and quite convinced that I had awakened from a long ontological sleep."

When Leary met Alpert at the airport in Mexico City shortly after, he greeted him, saying he had just been through six hours that taught him more than all his years studying psychology. "That was impressive to a fellow psychologist," Alpert remarked.

Concord Prison Project

The convicts Leary had were some of the toughest convicts in the Massachusetts prisons…. They were armed robbers. They ruled other convicts when the guards were out of sight. They had no compunctions about breaking somebody's arm, if that was necessary to enforce their ideas. They volunteered for this and thought they were going to get control of the experiment. Instead, these tough convicts all had profound religious experiences. One of the toughest of them told me about when he took psilocybin, he had a vision of Christ and he helped Christ carry his cross towards Calvary. Then he said that after the vision stopped, "I looked out of the window and all my life came before my eyes", and I said, "What a waste!"' Well, that was the turning point in this person's experiences. He and other tough guys started an organization within the walls to continue with their own rehabilitation and the rehabilitation of others.

—Walter Houston Clark

10
Similarity to LSD

LSD AND MESCALINE are generally thought to have more impact than psilocybin because of their longer duration. They are also perceived by many people as more coercive than psilocybin. The psilocybin experience seems to be warmer, not as forceful, and less isolating. It tends to build connections between people, who are generally much more in communication than when they use LSD.

Bad Trips

ALTHOUGH RARE, SOME "HELLISH" EXPERIENCES have resulted from psilocybin and mushroom use, especially when they were administered in inappropriate hospital settings by doctors unacquainted with their effects. A vivid account of one such trip appears in Ebin's The Drug Experience—a first-class example of how not to conduct such investigations.

Hofmann provides two examples of bad trips. The first occurred after swallowing thirty-two dried specimens of Psilocybe mexicana to see if Heim's cultivation from Mexican sources produced mushrooms that were still psychoactive.

The second appears as an experiment with psilocybin in his autobiography. Hofmann seems to be one of those people exquisitely sensitive to psychedelic effects. As with his first self-ingestion of LSD, these trips came on overwhelmingly. The first experiment involved a medium dose, by native standards, of 2.4 gm.

Hofmann's Bad Trip

Thirty minutes after taking the mushrooms the exterior world began to undergo a strange transformation. Everything assumed a Mexican character. As I was perfectly well aware that my knowledge of the Mexican origin of the mushrooms would lead me to imagine only Mexican scenery, I tried deliberately to look at my environment as I knew it normally. But all voluntary efforts to look at things in their customary forms and colors proved ineffective. Whether my eyes were closed or open I saw only Mexican motifs and colors.

When the doctor supervising the experiment bent over to check my blood pressure, he was transformed into an Aztec priest and I would not have been astonished if he had drawn an obsidian knife. In spite of the seriousness of the situation, it amused me to see how the Germanic face of my colleague had acquired a purely Indian expression.

At the peak of the intoxication, about one and a half hours after ingestion of the mushrooms, the rush of interior pictures,

mostly abstract motifs rapidly changing in
shape and color, reached such an alarming
degree that I feared that I would be torn
into this whirlpool of form and color and
dissolve. After about six hours the dream
came to an end. Subjectively, I had no idea
how long this condition had lasted. I felt
my return to everyday reality to be a happy
return from a strange, fantastic but quite
really experienced world into an old and
familiar home.

—Albert Hofmann

In Hofmann's 1962 psilocybin experiment,
undertaken with the novelist Ernst Junger, the
pharmacologist Heribert Konzett, and the Islamic
scholar Rudolf Gelpke, each took 20 mg. of psilocybin. Hofmann summarized the experience as
having "carried all four of us off, not into luminous
heights, rather into deeper regions" and concluded:
"It seems that the psilocybin inebriation is more
darkly colored in the majority of cases than the inebriation produced by LSD."

LSD and Interpersonal Isolation

IN CONTRAST, AT HARVARD THERE WERE NO BAD TRIPS on
psilocybin. Michael Kahn, a psychologist who observed both the Harvard and Millbrook psilocybin
scenes, gives an account of how the advent of LSD
changed the setting, resulting in greater emphasis
on solitary experience.

We Were On a "Love Trip"

There were no "bad trips" in those days.
We didn't know what a "bad trip" was.
Hundreds of psilocybin trips—I never saw

one. I didn't even have a word "bad trip" in my vocabulary. Those were benign, life-changing, growth experiences, because Tim's presence was so involving.... We were on a love trip; Timothy had us on a love trip and it was fantastic.

We just formed this incredible community. We saw each other every day and we hung around together, and we planned sessions together, and we played together, and we exchanged lovers, and it was just fantastic.

Then Michael came and introduced the LSD and some stuff happened that I really didn't like—I guess I should say I really didn't understand. Not so much to Timothy as the rest of us. LSD is a very different drug, and we began going on solo trips which we hadn't been doing so much. People would take these wild doses of LSD and disappear, and you wouldn't see them again for two days—including myself.

You know, you get together with the gang expecting one of those love sessions, and somebody would give you 400 kosher mcg. of that stuff and you'd never see anybody again till two days later, and you'd all look around and there you'd been out in the Tibetan mountains.

It was fun, and it was exciting, and it was scary. And the bad trips began—and the scary things. But then what happened that really disturbed me a lot was that we got quite cliquey—which had never happened before.

You see, things like this would happen: we
would finish the psilocybin session and
we would go out to the Dunkin' Donuts
or the Star Market to get breakfast. And
you never saw such a beautiful bunch of
people in your life as we were walking
into those places. Everybody else who was
waiting in line for the "Dunkin' Donuts"
were our brothers and sisters. We would
quick over to the end of the line, somebody
would come in and we would keep going
to the end of the line and we would strike
up conversations with these people. We
would have this far-out thing going with
the Dunkin' Donuts on Sunday morning,
you know.

LSD changed all that. We got snotty, we got
put-downy, we got "in" and "out." We got
looking at the people who hadn't had "the
experience" as though they were inferior to
us. We would go to parties and there would
be "drug people" and "non-drug people,"
and we would be in little groups, and we
would tell "in-jokes," and we would be
groupy, and we'd put down people who
tried to get in with us.

—Michael Kahn

Auditory Hallucinations

LSD AND MESCALINE HAVE A REPUTATION for being spec-
tacular hallucinatory. In moderate to high doses, psilocybin and psilocybian mushrooms produce striking
visual effects in most users who close their eyes, even
among people who are not "visualizers." In contrast
to most other psychedelics, psilocybian mushrooms
have impressed many with auditory effects.

Upbeat Mushroom Experience

The state of mind induced by a full dose
of mushrooms is one of euphoria and calm
lucidity, with no loss of coherence or clarity
of thought. The hallucinations seen with
the eyes closed are colorful, hard-edged,
and highly articulated, and may range from
abstract geometrical forms to visions of fantastic landscapes and architectural vistas.
These hallucinations are most intense when
the mushroom is taken in the setting pre-
ferred by the Mazatecans: inside at night
in complete darkness. On the other hand,
if one is in a natural setting and directs the
focus of the senses outward to the environment, one discovers that one's senses seem
keyed to their highest pitch of receptivity,
and finds oneself hearing, smelling, and
seeing things with a clarity and sensitivity
seldom, if ever, experienced before.

—Oss & Oeric

11
Wonderous Trips

HEN HOFFER AND OSMOND TOOK UP THE MAT-
TER of mental effects in their book The
Hallucinogens, they remarked that "the
major difference between the mushroom effect and pure psilocybin seems
to be the dryness of the scientific accounts and the
richness of the accounts of self-experimentation."
Probably no finer example of richness exists than
in the descriptions of Wasson. In The Wondrous
Mushroom he wrote about four psilocybian experiences, highlighting the contributions of mental
set and setting. All four experiences involved the
same mushroom—Psilocybe caerulescens—taken in
roughly the same dosage.

Wasson's Trips

AS TO MENTAL SET, Wasson's expectations had been
building for some years. While his father was interested in religion and drink, Wasson became steadi-
ly more interested in religion and mushrooms,
studying their significance in various cultures for
a quarter of a century before he acquired a sample
of teonanácatl. Another two years passed before he
actually tasted it, as he waited to find someone who
could perform the mushroom ceremony.

Trip 1

ON THE VERGE OF A MUCH ANTICIPATED but unfamiliar experience, Wasson derived an important sense of reassurance from his setting. He was in the company of Indians who were experienced users and who were taking the mushroom with him. The velada was conducted by a sabia—a wise-woman, "one-who-knows." Wasson had a close friend along for a companion.

The ceremony consisted of chanting that continued all night long, except for brief intermissions every forty minutes or so. The sabia Maria Sabina danced for two hours in the dark. The ritual aspect, in the context of feeling both adventurous and safe, seems to have influenced the quality, or tone, of Wasson's experience.

Visions

The visions came in endless succession,
each growing out of the preceding one.
We had the sensation that the walls of our
humble house had vanished, that our un-
trammeled souls were floating in the empy-
rean, stroked by divine breezes, possessed
of a divine mobility that would transport
us anywhere on the wings of a thought.
—R. Gordon Wasson

Trip 2

THAT FIRST EXPERIENCE WAS IMPRESSIVE—even "gala." Wasson and his associate Allan Richardson had many questions to clarify. Three days later, they asked if Maria Sabina would perform a second velada. This time Richardson didn't ingest any mush-

rooms because he intended to take photographs.
Wasson accepted five pairs of the "landslide" or
derrumbe mushrooms, rather than the six pairs he
had taken on the first trip.

Wasson said the effects were just as strong. He
felt nauseous the first time and twice had to leave
the room. On this occasion, he didn't have that
problem. The setting was the same, but his mental
set was quite different.

Same Setting, Different Mental Set

It was raining in torrents all that night, so
there was no moon. But the Señora's behavior differed much from what we had seen
the first time. Every thing was reduced in
scale. There was no dancing and virtually
no percussive utterances. Only three or
four other Indians were with us, and the
Señora brought with her, not her daughter, but her son Aurelio, a youth in his late
teens who seemed to us in some way ill
or defective. He was now the object of her
attention, not I. All night long her singing
and her words were directed to this poor
boy. Her performance was the dramatic
expression of a mother's love for her child,
an anguished thronody to mother love, and
interpreted in this way it was profoundly
moving. The tenderness in her voice as she
sang and spoke, and in her gestures as she
leaned over Aurelio to caress him, moved
us profoundly.

—R. Gordon Wasson

Trip 3

THREE DAYS LATER, Wasson's wife Valentina Pavlovna ingested five pairs and their thirteen-year-old
daughter Masha took four pairs of the same mushrooms. They swallowed them during the afternoon in sleeping bags in a closed room. It was the
first occasion, Wasson remarked, "on which white
people were eating the mushrooms experimentally,
without the setting of a native ceremony. They too
saw visions, for hours on end, all pleasant, mostly
of a nostalgic kind. VPW at one point thought she
was looking down into the mouth of a vase, and
there she saw and heard a stately dance, a minuet,
as though in a regal court of the 17th Century. The
dancers were in miniature and the music was oh!
so remote, but also so clearly heard. VPW smoked
a cigarette; she exclaimed that never before had a
cigarette smelled so good. It was beyond earthly
experience. She drank water, and it was superior to
Mumm's champagne—incomparably superior...."

Trip 4

SIX WEEKS LATER, Wasson experienced the mushrooms in New York. Although dried, they apparently retained much of their potency. He wondered subjectively "if indeed their power had not
increased." Secure in his home and confident now
about the mushrooms, Wasson found the setting
was actually enhanced by a terrific storm, Hur-
ricane Connie. "As I stood at the window and
watched the gale tossing the trees and the water of
the East River, with the rain driven in squalls before
the wind, the whole scene was further quickened
to life by the abnormal intensity of the colors that I
saw. I had always thought that El Greco's apocalyp-

tic skies over Toledo were a figment of the painter's
imagination. But on this night I saw El Greco's
skies, nothing dimmed, whirling over New York."

Trips Analyzed

FOUR EXPERIENCES, catalyzed by the same mushroom,
yielded four considerably different results. Two years
later, Wasson amalgamated these into his generalized
description of the effects of psilocybian consumption.

Different Effect with Different People

The mushrooms take effect differently
with different persons. For example, some
seem to experience only a divine euphoria,
which may translate itself into uncontrolla-
ble laughter. In my own case I experienced
hallucinations. What I was seeing was
more clearly seen than anything I had seen
before. At last I was seeing with the eye of
the soul, not through the coarse lenses of
my natural eyes.

Moreover, what I was seeing was impreg-
nated with weighty meaning: I was awe-
struck. My visions, which never repeated
themselves, were of nothing seen in this
world: no motor cars, no cities with sky-
scrapers, no jet engines. All my visions
possessed a pristine quality: when I saw
choir stalls in a Renaissance cathedral, they
were not black with age and incense, but
as though they had just come, fresh carved,
from the hand of the Master.

The palaces, gardens, seascapes, and mountains that I saw had that aspect of newness,

of fresh beauty, that occasionally comes to
all of us in a flash. I saw few persons, and
then usually at a great distance, but once I
saw a human figure near at hand, a woman larger than normal, staring out over a
twilight sea from her cabin on the shore.
It is a curious sensation: with the speed of
thought you are translated wherever you
desire to be, and you are there, a disembod-
ied eye, poised in space, seeing, not seen,
invisible, incorporeal.

—R. Gordon Wasson

Andrew Weil's Trips

IN THE MARRIAGE OF THE SUN AND MOON, Andrew
Weil declared that he's a mycophile and described
three trips he took using the San Isidro mushroom
called Stropharia cubensis. These three experiences,
with the same mushroom, stimulated greatly varying responses. They emphasize again the significant
influence of mental set and setting.

Trip 1

WEIL ARRIVED IN HUAUT-
LA DE JIMÉNEZ, where he
had the good fortune to
be taken into the house
of a curandera living in
a nearby village. As a
healer, she used mod-
ern medicines and also
mushrooms, which she
regarded as the gran
remedio that cures all
ills. She had already col-

Dr. Andrew Weil and John W. Allen, an expert in entheogenic

lected a bunch of San Isidro mushrooms that were
obviously meant for Weil.

> There were larvae and insects among the
> mushrooms. The curandera, however,
> passed the mushrooms through the smoke
> from a dried chile pod placed on glowing
> charcoal, and instantly the insects crawled
> out of the mushrooms and died on a news-
> paper placed below. Weil ate two of the
> largest mushrooms, three-inch caps. As the
> curandera prayed, he ate twenty smaller
> ones.

Protected by the sacred ministrations of the
curandera, Weil was soon feeling "extraordinari-
ly content and well" and experienced sensations
of lightness. He felt "fresh, alert, healthy, and
cleansed." The healer communicated "much of

her own vitality, optimism, and goodness of spirit, leaving me elated and more confident in my own abilities and powers."

Going outside later, Weil recalled Wasson's suggestion that the word bemushroomed would be a good term for this state. He observed a full eclipse of the moon and later went to sleep. "In the morning, I awoke refreshed, feeling better than I had in a long time, and went off for a day in Huautla of shopping and negotiating with the military authorities...."

Trip 2

WHEN WEIL RETURNED, the healer told him that some mushrooms were left over and that he might as well finish them that night. "I really did not want to," writes Weil, "since I had just had a perfect mushroom experience, but instead of telling her that, I agreed." They repeated the service with incense and prayers beneath a picture of the patron saint Isidro, who was being showered with "psychedelic rays...from some other dimension."

A Different Experience

A heavy bank of fog and clouds closed in, the temperature dropped, and suddenly nearly everyone in the house was sick. There was much crying and coughing from the bedroom, and I began feeling unwell, too. A great sense of depression and isolation came over me. I could not get to sleep. The mushrooms seemed to be working against me, not with me, and I felt far away from where I was supposed to be.

—Andrew Weil

Toward dawn, Weil was still awake and con-
cluding that mushrooms, like other psychedelics,
"must be used in a proper context." He comments
on this lesson:

My Lesson

To take them just because they are available, when the time is not right, is a mistake. The negative experience of this second night did not in any way detract from
the goodness of the first night. If anything,
it made me more aware of the value of that
experience and more eager to retain it and
use it in my life. I hoped that I would be
able to be bemushroomed again, but I re-
solved to be patient until the right moment
came.

—Andrew Weil

Trip 3

A SHORT WHILE LATER, outside Cali in Colombia,
Weil ate Stropharia cubensis again. The mushroom
seemed to be growing all over the place, although
its use was not traditional there. Whites and others
"introduced Colombian Indians to the drug, the re-
verse of the usual order of things." The setting this
time was "an idyllically beautiful field with clumps
of woods, a clear river, and enormous, gray, hump-
backed Brahma cows lying peacefully in the bright
green grass."

Deeper Look into Setting

We sat in the grass, about ten of us, and let
the mushrooms transport us to a realm of
calm good feeling in which we drank in

the beauty of the setting. There were color visions, as I had experienced before with San Isidro in Mexico.

In Mexico I had eaten the mushrooms late at night, in darkness and secrecy, in the very shadow of menacing police authority. Now it was broad daylight, in open country, with no one around but friendly fellow travelers. In Mexico I had felt like an early Christian pursuing the sacrament in a catacomb, wary of the approach of Roman legions; here everything was aboveground and open. The Indians of the Sierra Mazateca say the mushroom should not be eaten in daytime, that they must be eaten at night. Yet here we were in full daylight, having a wonderful time.

In general, I prefer to take psychedelic substances in the daytime, when their stimulating energies are more in harmony with the rhythms of my body. I feel that way about mushrooms, too. Is it possible, I wondered, that the Indian habit of eating mushrooms at night is not so traditional as it seems, but dates back only to the arrival of the Spanish and persecutions of native rites by the church?

—Andrew Weil

Weil has since experimented with other psilocybian mushrooms. He feels that "the most interesting properties of mushrooms may not come to our attention if people use them casually and without thought."

12
Cubensis

HE PSILOCYBE CUBENSIS mushroom was originally collected in Cuba in 1904 and it is the easiest of all mushrooms to grow—even easier than the commercial ones. It is considered the best psilocybian mushroom for most users, and is certainly the most readily available. Its psychoactivity varies, however, and degenerates over time, especially when there is delay in moving the product from the grower to user. Some dealers have been known to open up bags of these mushrooms so they will gain more weight from ambient humidity, which breaks down the psilocybin and psilocin even further.

Psilocybe cubensis was collected by Schultes during his 1939 trip to Oaxaca and deposited at the Farlow Herbarium at Harvard University, while the mycologist Rolf Singer worked there identifying mushrooms. In an attempt to reorganize the taxonomy of the genus Stropharia, Singer came upon Schultes' specimens. In 1951, he placed Strophoria cubensis Earle in the Psilocybe genus, basing this identification on microscopic characteristics, particularly of the spore. He neglected to tell Schultes and didn't follow this work up, but the mushroom is now often referred to as Psilocybe cubensis (Earle) Singer.

Dispute

SOME MYCOLOGISTS HAVE DISPUTED Singer's identification, arguing that the macroscopic features place this mushroom in the genus Stropharia. Both designations are found in the literature—Stropharia in ethnopharmacological and some European mycological sources, Psilocybe mainly in the American, Mexican, and botanical sources.

Psilocybe cubensis is tropical and subtropical and appears in cowfields during rainy seasons or other times of high humidity. In the U.S., it is distributed mainly along the southeastern seaboard, but it can be found inland as far north as Tennessee.

Low Status

PSILOCYBE CUBENSIS GROWS NATURALLY in connection with cattle—particularly hot-weather-loving Brahmas—and especially in dung a few days old. It may be because of this that the Indians consider it inferior, and only use it as a last resort. More likely, it has lower status because it wasn't indigenous to Mexico. It arrived with Spanish importation of Brahma cattle from the Philippine Islands, and, thus, doesn't have ancient associations in indigenous shamanistic rites.

In Mexico, the Mazatecs call the Psilocybe cubensis species di-shi-tho-le-rra-ja, meaning "sacred mushroom of cow dung." Its other names are Spanish, such as San Isidro Labrador or "St. Isidore the Plowman." Anthropologist Peter Furst promoted the notion that it may have grown in deer dung, but attempts to grow it in that medium have not succeeded. In cultivating Psilocybe cubensis, dung isn't essential because it fruits very potently on rye

and other grains. Rye-cultured specimens appear
less robust than those in the field or those grown on
compost. In terms of psychoactivity they are about
a third stronger on the average.

Habitat

IN THE FIELDS, PSILOCYBE CUBENSIS MUSHROOM TENDS to
come up singly or in small groups. Growth is rapid.
In pastures, it often grows from the size of a pin-
head to a full-sized mushroom in little more than
a day. Psilocybe cubensis becomes rather large,

Stropharia (Psilocybe) cubensis.

Jeremy Bigwood

generally attaining a height of 15-30 cm. or 6 1/2 inches. It often appears whitish overall, sometimes with a streaking of "comic strip blue."

Psilocybe cubensis often blues without any apparent bruising, perhaps because of intense heat. When it opens, the cap usually gets lighter on the outside while the center gets darker—but

Stropharia (Psilocybe) cubensis, the most popular psilocybian mushroom—yet, the weakest by weight—is commonly illicitly grown throughout the world. These beauties were photographed on Koh Samui Island in the gulf of Thailand.

not necessarily so. It can be light brown or light reddish-gold, and often the prominent annulus, or collar, is covered with spores.

Psilocybin Content

EARLY STUDIES MADE OF PSILOCYBE CUBENSIS estimated the concentrations of psilocybin at about 0.2 percent (2.0 mg./gm.) of the dry weight, along with a fairly high amount of psilocin. The analytical tools then in use necessitated heating and are considered obsolete. State-of-the-art equipment—such as High-Performance Liquid Chromotography (HPLC) shows that much of the psilocin observed was actually psilocybin that had been transformed into psilocin by the analytical process.

Bigwood and Beug reported psilocybin in concentrations as large as 13.3 mg./gm. (1.3 percent) and psilocin in concentrations of 1.0 mg./gm. in a batch of dried Stropharia cubensis mushrooms. In the same strain, however, they also found psilocybin as low as 3.2 mg./gm. and psilocin at 1.8 mg./gm.

Moreover, they discovered that the same strain in the same container produced greatly varying amounts of psychoactive constituents in different flushes appearing about a week apart. For example, they recorded the following psilocybin content—in mg./gm. dry weight—for one sequence of five flushes: 8.3, 6.5, 13.3, 4.8, and 6.8. The potency of the third flush was twice that of the second and nearly three times that of the fourth.

The third flush did not show the highest psychoactive concentrations in other instances, however. The only consistency found was that in the first

flush psilocin—and the psilocybin and psilocin analogs—were either barely present or entirely absent.
Their strength then increased in subsequent flushes.

Conclusions

We found that the level of psilocybin and
psilocin varies by a factor of four among
various cultures of Psilocybe cubensis
grown under rigidly controlled conditions,
while specimens from outside sources
[street samples] varied ten-fold. It is clear
that entheogenic and recreational users
of this species have no way to predict the

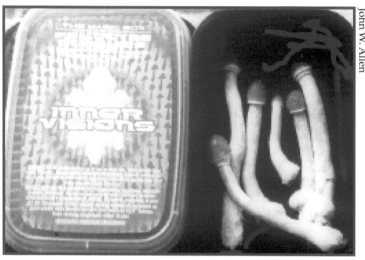

*Often packaged and freeze dried for later use or sales, both Holland
and the UK now offer legal mushrooms for sale commercially in
shops and over the internet. Above is a one- ounce dosage (equal
to 3-5 dried grams, a clinical dosage) for sale in Amsterdam at
the Conscious Dreams Collective Smart Shop. One of five shops
operated by Hans Van Den Hurk, the first person to legally sell fresh
mushrooms in the Netherlands.*

of this species have no way to predict the amount of psilocybin and psilocin they are ingesting with a given dry weight of the mushroom. Thus, variations in the subjective experience come not only from effects of 'set' and 'setting,' but also in very real measure from dosage differences.
                        —Jeremy Bigwood & Michael Beug

13
Liberty Caps

PSILOCYBE SEMILANCEATA IS THE SECOND most important psilocybian mushroom worldwide. Its popular name—Liberty Caps—comes from its resemblance to caps worn during the French Revolution. Found in the northern temperate zone, Liberty Caps grow inland, up to a thousand miles from the ocean. It fruits in tall grasses and on cow fields mainly in the fall, and only occasionally in the spring. Unlike Psilocybe cubensis, it does not grow on dung itself. This species is fairly small, about 10 cm. or 4 inches tall, at most.

Early Reports

REPORTS OF THE PSYCHOACTIVITY of Liberty Caps predate the Wassons' journeys to Mexico. In McIlvaine's One Thousand American Fungi, this species is described as a mushroom with strange effects that don't last long and are not toxic. The psychoactive ingredient, according to this classic text, can be removed by boiling the mushroom in water and then throwing away the water. A professor at Yale and his wife took it a few times in 1910 and had marvelous experiences and hilarity for a short while. At about this same time, reports circulated of its use by the artist community in Norway, Maine.

Peter Stafford

J.Q. Jacobs / Hallucinogenic & Poisonous Mushroom Field Guide

Psilocybe semilanceata, better known as Libery Caps.

PSILOCYBIN WAS FIRST DETECTED in Psilocybe semilanceata by Heim and Hofmann in the early 1960s,
but it was not used in Europe until at least a dozen
years later. In the Pacific Northwest, use began as
early as 1965. The Royal Canadian Mounted Police
confiscated some on the Vancouver campus of the
University of British Columbia at that time. Experimentation with Psilocybe semilanceata was only
sporadic, however, until publication of Enos' book.
Today thousands of people in the northwestern
U.S., Scandinavia, especially Norway, the British
Isles, and most of western Europe collect this mushroom.

*Psilocybe pelliculosa is a relatively weak species macro-scopically
similar to Psilocybe semilanceata—the liberty cap.*

Psilocybin Content

PSILOCYBE SEMILANCEATA IS QUITE POTENT by weight,
containing as much as 12.8 mg. psilocybin/gm. and
averaging around 11 mg./gm. in dried specimens.
Unlike Psilocybe cubensis, which is highly variable

in strength, Psilocybe semilanceata is much more uniform with psychoactivity differing in samples by not much more than a factor of two. Three or four of these tiny mushrooms are often enough to energize the body and affect color perception and vi- sual acuity. About twenty to thirty mushrooms con- stitute a strong dose, although some people have been known to take up to 100 at a time. It is recommended no more than 2 or 3 dried grams be taken the first few times, however.

Some people have observed that mushrooms containing both psilocy- bin and psilocin tend to lose their psychoactivity fairly rapidly, whereas those lacking psilocin tend to have a long shelf life. Liberty Caps have no psilocin, but they do contain psilocybin ana- logs. Specimens once were reanalyzed after four years on a shelf without refriger- ation, and were found to be almost as potent.

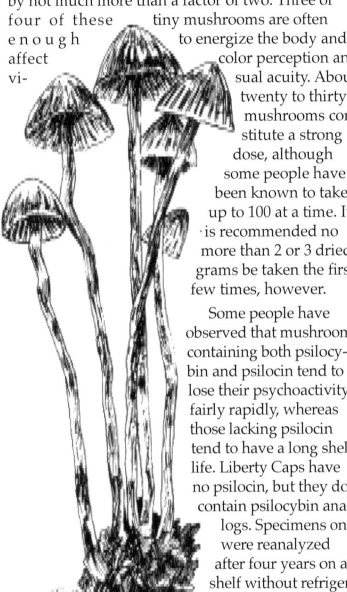

Psilocybe pelliculosa

Confusing

MOST LIBERTY CAPS DO NOT BLUE, although some do so heavily. It is not known whether the latter are another species. Some mushrooms are quite similar in appearance to Psilocybe semilanceata and inhabit the same cow fields. These may be variants or other new species. Of those that look like Liberty Caps and grow on pastures, none is poisonous. There is a psilocybian mushroom that is similar and is known as Psilocybe linaformens, common in Europe and Oregon, where it too is often called a Liberty Cap.

Psilocybe pelliculosa, another psilocybian species that many people confuse with Liberty Caps, contains psilocybin and the same analogues but has only half the potency of Liberty Caps. This one can be easily distinguished by habitat. Rather than in cow fields, it grows on sawdust or wood chip piles in forested areas where lumberjacks have been working. It doesn't display the indrawn edge to its cap that is evident in Liberty Caps, and often its stem and cap edge darken with age.

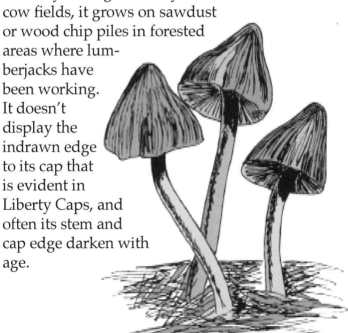

14
Benanosis

ANAEOLUS SUBBALTEATUS is the most prevalent of the psilocybian mushrooms, growing throughout the U.S. and found in various climates in many parts of the world. It grows singly or in clusters to a height of about 8 cm.—just over 3 inches—most commonly on composted dung, in both spring and fall. The tan cap develops a striking cinnamon-brown band around the bottom and flattens as it ages, with the central portion fading over time to a pale, warm buff color. Eventually Panaeolus subbalteatus resembles a large floppy hat draped over a fairly thick, whitish stem that sometimes blues at its bottom. The spore print has a blackish-purple color.

Panaeolus subbalteatus springs up on compost, straw, and manure piles and often can be seen clustered in one- or two-foot rings along roads, on lawns, and in open areas. It can be cultivated, but only on compost.

Mixed Opinions

ANDY WEIL CLAIMS THAT Panaeolus subbalteatus doesn't produce as good a psychedelic trip as most other psilocybian mushrooms and may bring on

Panaeolus subbalteatus grows world-wide and is commonly found in rotting hay stacks, cow manure, and compost heaps at riding stables and race tracks in the early spring and early fall.

stomach aches. These assertions have not been confirmed by most other users. Many people like Panaeolus subbalteatus—especially in the spring when few Liberty Caps are available.

Psilocybin Content

PANAEOLUS SUBBALTEATUS HAS A LONG shelf life and contains no psilocin. The psychoactivity comes only from psilocybin and its analogs. The amounts are low to moderate for psilocybin, varying from a little over 1.5 mg./gm. up to 6.0 mg./gm. dry weight.

Michael B. Smith / Halluncinogenic & Poisionous Mushroom Field Guide

Panaeolus subbalteatus

15
Wavy Caps

SILOCYBE CYANESCENS is the most potent psilocybian mushroom known. Although not as big as Psilocybe cubensis, it is probably the second largest growing in the U.S., generally reaching 6-8 cm. in height, which is about 3 inches. This mushroom, like Psilocybe pelliculosa and the few others that follow, don't grow on dung but rather on hardwoods and wood chips.

Habitat

PSILOCYBE CYANESCENS LIKES TO INHABIT landscaped yards containing ground bark and often dwells under Douglas fir or cedar and in mulched rhododendron beds. The large cap, which starts out chestnut brown and becomes more caramel-colored with age, is wavy, so they've been called "Wavy Caps."

Because of the blue line around the edge of the cap, they are also known as "Blue Halos." Generally this species grows in clusters, although sometimes it comes up singly. When it fruits, it is prolific. It is not unusual to hear of mushroom hunters gathering six to seven ounces dry weight at a time.

Teonanacatl

Psilocybe cyanescens

Psilocybin Content

WAVY CAPS CONTAIN psilocybin, psilocin, and at least four analogs. It blues markedly when bruised and is less stable than Liberty Caps. It isn't cultivated very easily, but it is potent. Bigwood and Beug say that, "Psilocybin levels were found to range up to 16.8 mg./gm. dry weight, with the total psilocin plus psilocybin levels reaching 19.6 mg./gm. dry weight—nearly 2% of the dry weight." Being large and the most potent psilocybian species, it is quite popular and one of the most sought after mushrooms in the Pacific Northwest.

John W. Allen

Psilocybe cyanescens is a large species with an undulating caramel-colored chestnut cap. It is the most potent psilocybian mushroom found anywhere. It fruits prolifically from fall through winter from San Francisco to British Columbia, Canada.

16
Baeocystis

SILOCYBE BAEOCYSTIS is another strongly bluing Pacific Northwest species. It can be found growing on ground bark, wood chips, peat moss, and occasionally on lawns. This popular mushroom appears from fall through midsummer. As many as fifty are often clumped together. Generally it contains rather low levels of psilocybin and psilocin. Its potency, however, is highly variable.

Identification

THE EDGE OF THE CAP of the Psilocybe baeocystis generally undulates, resembling a bottle cap or lawn umbrella, and the stem is often characterized by twisting bends. An important identifying characteristic is a brown spot that appears in the center of the cap after it is dried. Psilocybe baeocystis is bound to become more familiar because of increasingly widespread use of its host materials in industrial parks and around homes.

Another psilocybian mushroom that also blues and has characteristics similar to Psilocybe baeocystis and Psilocybe cyanescens is Psilocybe cyanafiberlosa. It was originally collected between patches of the first two. Psilocybe cyanafiberlosa has somewhat wavy caps like Psilocybe cyanescens, although they don't open up nearly so much, and the same chemical components as the others. It's normally weaker than Psilocybe baeocystis, often containing about 1.3 mg. of both psilocybin and psilocin/gm. dry weight.

Psilocybin Content

BIGWOOD AND BEUG ASSESS Psilocybe baeocystis as averaging about 2.8 mg. psilocybin/gm. and 1.4 mg. psilocin/gm. dry weight. But one collection they examined was found to contain 8.5 mg. psilocybin/gm. and 5.9 mg. psilocin/gm.—rivaling concentrations in the strongest psilocybian species. This variability makes it riskier than most others to use. "One could eat a lot of weak ones to get an experience," comments Bigwood, "and then go to another patch and get maybe 200 mg. of psilocybin plus."

Psilocybe baeocystis

John W. Allen

Psilocybe baeocystis, like P. cyanescens, is a woodchip and bark mulch mushroom in the Pacific Northwest, often fruiting in both lawns and mulched garden beds. It is a potent species in small amounts but relatively weak in its dried stage.

17
Stuntzii

SILOCYBE STUNTZII is another psilocybian mushroom that readily springs up on commercially prepared wood chips. It is a fairly small, stout mushroom, which in many people's opinion should be included in the Stropharia genus since it has a prominent annulus. Psilocybe stuntzii can be found on grasses, bark mulch, and mulched lawns in the fall and summer. Under the grasses there is always a layer of wood bits. The cap is sticky and has stripes on the edges that become flat, scalloped edges when the mushroom is mature.

This is very likely the weakest psilocybian mushroom in the Pacific Northwest. The Bigwood and Beug studies show that its potency ranges from zero to a high of 3.6 mg. psilocybin/gm. dry weight.

Identification

CAUTION IS IN ORDER WHEN GATHERING Psilocybe stuntzii. It can easily be confused by the novice with the deadly species Galerina autumnalis, with which it sometimes clusters. The most potent psilocybian mushrooms grow on wood chips, and so anyone hunting for them should know how to identify the poisonous Galerina.

Psilocybe stuntzii growing beside Galerina autumnalis.

Psilocybe stuntzii and Galerina autumnalis can definitely be separated from one another by spore color. Spores of the Galerina species are rust-brown in contrast to those of the Psilocybe species, which are gray to lilac. Furthermore, their coloration is different, their collars look quite different, and the Galerina species presents lines radiating from the center of the cap. Only the psilocybian mushroom turns blue upon bruising. If there is any doubt as to identification, a mushroom that does not blue should be discarded as Galerina. The toxic ingredient in Galerina autumnalis also appears in the common edible mushroom, but in very small amounts.

Risky

ON WHIDBY ISLAND IN WASHINGTON STATE, some teenagers searching for psilocybian mushrooms made a mistake in identification. Two young men and a woman who ingested the mushrooms got sick the next day, but were afraid to report their illness for fear of arrest by the authorities. They apparently waited yet another day before the symptoms got extremely bad, and then all three went to a hospital. The two young men survived, but the young woman died on Christmas Day.

Another fatality attributed to the search for psilocybian mushrooms occurred when two adults and four children ate a large number of what they as-

Psilocybe stuntzii

John W. Allen

*A composite of three collections of Psilocybe stuntzii,
another Pacific Northwest species which fruits abundantly in public
areas, in both lawns and mulched gardens, sometimes fruiting all
year long in warmer conditions. Also known as "blue ringers"
because of the ring on the stem , a veil remnant damaged when the
cap opens to release the spores—thus causing the ring to stain blue.*

sumed were Psilocybe baeocystis. A six-year-old boy in the group died. Some of the mushrooms from this patch, photographed for an article, were clearly not Psilocybe baeocystis but Psilocybe cyanescens. Because these two species sometimes grow together with Galerina autumnalis, it's likely that the six people ingested all three.

> There were some mycologists who stated publicly that it would be better for people to die from mistakes in identification than to provide them with the tools for recognizing a Psilocybe mushroom.
> —Paul Stamets

Consult an Expert

THIS SHORT DESCRIPTION of the main North American species illustrates how varied psilocybian mushrooms are. People intending to gather or cultivate them should consult experts, particularly when identification of a mushroom is in question. Even clear photographs may be only somewhat helpful. It should be kept in mind that mushrooms change appearance as they age and often have different coloration in different regions.

18
Soma

GORDON WASSON SHOOK UP entheobotanists when he asserted that the Amanita muscaria mushroom—known also as Fly Agaric—fits the criteria for the mysterious, lost Soma. Wasson drew much of his evidence from the Rig-Veda—a collection of 1,028 religious hymns written by the Aryans dating back to the second millenium B.C. One hundred twenty of the verses from the first of the Vedas, which has been credited as being the foundation of modern Hinduism, are devoted to praise of a plant called Soma, which is characterized as being rootless, leafless, blossomless, and from the mountains.

SOMA
DIVINE MUSHROOM
OF IMMORTALITY
by
R. Gordon Wasson

Soma: The Devine Mushroom of Immortality by R. Gordon Wasson.

Fly Agaric

AMANITA MUSCARIA—OR FLY AGARIC as it is often called—is a red mushroom speckled with white, often found in drawings accompanying fairy tales. In a his book SOMA: Divine Mushroom of Immortality, Wasson explored the Rig-Veda's evidence and his thesis that Soma was actually the Fly Agaric mushroom. Authorities such as Schultes and Hofmann have concurred.

Ray Rogers / Blotter #3

Keewaydinoquay

Wasson explained that "as I entered into the extraordinary world of the Rig-Veda, a suspicion gradually came over me, a suspicion that grew into a conviction: I recognised the plant that had enraptured the poets…. As I went on to the end, as I immersed myself ever deeper in the world of Vedic mythology, further evidence seeming to support my idea kept accumulating. By Jove, I said, this is familiar territory!"

Wasson presented his "surprising new discovery" at a mushroom conference in San Francisco in the late 1970s, where he discussed evidence that the Fly Agaric mushroom had been used extensively by Indian tribes around the Great Lakes and eastward. The audience was thrilled by the introduction of Keewaydinoquay, a lively Ojibway Indian woman and university-trained ethnobotanist in her sixties, who has been ingesting the mushrooms three to five times a year since the age of fourteen.

Grows Under Trees

FLY AGARIC AND PANTHER CAPS grow only in mycorrhizal relationship with just a few trees—the birch, larch, fir, pine, and oak. A symbiotic association between the root cells of these trees—living or dead—and the fungus' mass of underground filaments is necessary for the mushroom to sprout. This particularity about growth conditions helped Wasson explain why the Soma of the Rig-Veda got "lost." For the most part India lacks forests of birch, fir, pine, and oak.

Wasson theorized that the Aryans found the mushroom growing in the Himalayas. However, as they lost contact with the mushroom, it's not at all surprising that the later parts of the Rig-Veda speak of "soma substitutes." The fact that Amanita muscaria grows only in a mycorrhizal relationship may also explain, according to Wasson, why the birch is held in such high regard in many northern lands.

In Northern California, especialy around Big Sur, the Fly Agaric appears in the bright red coloration familiar from countless children's books. In other parts of North America, its color varies from pink and even white to bright canary yellow. Similar variations occur in Europe and Asia.

Amanita muscaria, the red, spotted Fly Agaric mushroom.

Jeremy Bigwood

Native American Practices

A LETTER DATED 1626 FROM A JESUIT in Quebec to his brother in France has surfaced. The Jesuit described the North American Indian practices a full century before any published references to Siberian mushroom practices.

"They assure you that after death they go to heaven where they eat mushrooms and hold intercourse with each other."

Fascinated, Wasson compared Native American practices with those in Siberia and concluded that the rituals were similar and that shamanistic employment of Fly Agaric was "circumpolar in extent." In their book, Mushrooms, Russia and History, the Wassons chronicled the remarkable similarities in practices and beliefs. The only important difference was in regard to the "reindeer symbolism" associated with the mushrooms in Siberia. In North America, there is no such symbolism because there are no reindeer.

The Wassons comment: "With our Mexican experiences [with psilocybian mushrooms] fresh in mind, we reread what Jochelson and Bogoras had written about the Korjaks and the Chukchees. We discovered startling parallels between the use of the fly amanita (Amanita muscaria) in Siberia and the divine mushrooms in Middle America. In Mexico the mushroom 'speaks' to the eater; in Siberia 'the spirits of the mushrooms' speak. Just as in Mexico, Jochelson says that among the Korjaks 'the agaric would tell every man, even if he were not a shaman, what ailed him when he was sick, or explain a dream to him, or show him the upper world or the underground world or foretell what would happen to him.' Just as in Mexico on the following day those who have taken the mushrooms compare their experiences, so in Siberia, according to Jochelson, the Korjaks, 'when the intoxication has passed, told whither the fly-agaric men had taken them and what they had seen.' In Bogoras we discover a link between the lightning bolt and the mushroom. According to a Chukchee myth, lightning is a One-Sided Man who drags his sister along by her foot. As she bumps along the floor of heaven, the noise of her bumping makes the thunder. Her urine is the rain, and she is possessed by the spirits of the fly amanita…."

Siberian Tribal Use

THE EARLIEST REPORT FOUND BY the Wassons about Siberian Amanita muscaria practices came from a Polish prisoner of war, who wrote in 1658 about the Ob-Ugrian Ostyak of the Irtysh region in western Siberia. It says: "They eat certain fungi in the shape of fly-agarics, and thus they get drunk worse than on vodka, and for them that's the very best banquet."

Michael S. Smith
Hallucinogenic & Poisonous Mushroom Field Guide

The Fly Agaric mushroom — Amanita muscaria.

The first published account of Fly Agaric appeared in 1730, in the work of a Swedish colonel
who spent twelve years as a prisoner in Siberia. He
indicated that the Koryak tribe would buy a mushroom "called, in the Russian Tongue, Muchumor,"
from Russians in exchange for "Squirils, Fox, Her-
min, Sable, and other Furs…. Those who are rich
among them lay up large Provisions of these Mushrooms for the Winter. When they make a Feast,
they pour Water upon some of these Mushrooms,
and boil them. They then drink the Liquor, which
intoxicates them…. Of this Liquor, they…drink so
immoderately, that they will be quite intoxicated, or
drunk with it."

The tribesmen in Siberia did not know about
alcohol until after contact with the Russians. Johann
Georgi, in a book on Russia published in German in
St. Petersburg in 1776, remarked on the difference:
"Numbers of the Siberians have a way of intoxicating themselves by the use of mushrooms, especially
the Ostyaks who dwell about Narym. To that end
they either eat one of these mushrooms quite fresh,
or perhaps drink the decoction of three of them.
The effect shows itself immediately by sallies of wit
and humour, which by slow degrees arises to such
an extravagant height of gaiety, that they begin to
sing, dance, jump about, and vociferate: they compose amorous sonnets, heroic verses, and hunting
songs. This drunkenness has the peculiar quality
of making them uncommonly strong; but no soon-
er is it over than they remember nothing that has
passed. After twelve or sixteen hours of this enjoyment they fall asleep, and, on waking, find themselves very low-spirited from the extraordinary
tension of the nerves: however, they feel much less

head-ache after this method of intoxication than is produced by spiritous liquors; nor is the use of it followed by any dangerous consequences."

Few Accounts

THE EARLIEST REPORT FROM SOMEONE who had actually eaten a Fly Agaric mushroom appeared in 1837, in Polish. In 1797, ill and running a fever, Joseph Kopéc was given a mushroom as medicine by an evangelist, who first told him that "Before I give you the medicine I must tell you something important. You have lived for two years in Lower

A fresh specimen of Amanita muscaria, the red-spotted "fly Agaric" familiar from fairy tales, which grows in association with Birch and

Kamchatka but you have known nothing of the treasures of this land. Here are mushrooms that are, I can say, miraculous. They grow only on a single high mountain close to the volcano and they are the most precious creations of nature."

Kopéc ate half a mushroom and almost immediately he went into a deep sleep: "I found myself as though magnetized by the most attractive gardens where only pleasure and beauty seemed to rule. Flowers of different colours and shapes and odours appeared before my eyes; a group of most beautiful women dressed in white going to and fro seemed to be occupied with the hospitality of this earthly paradise. As if pleased with my coming, they offered me different fruits, berries, and flowers. This delight lasted during my whole sleep, which was a couple of hours longer than my usual rest...."

Kopéc ate the second half of the mushroom. "For several hours new visions carried me to another world, and it seemed to me that I was ordered to return to earth so that a priest could take my confession. This impression, although in sleep, was so strong that I awoke and asked for my evangelist. It was precisely at the hour of midnight and the priest, ever eager to render spiritual services, at once took his stole and heard my confession with a joy that he did not hide from me. About an hour after the confession I fell asleep anew and I did not wake up for twenty-four hours. It is difficult, almost impossible, to describe the visions I had in such a long sleep; and besides there are other reasons that make me reluctant to do so.

"What I noticed in these visions and what I passed through are things that I would never imagine even in my thoughts. I can only mention that

from the period when I was first aware of the no-
tions of life, all that I had seen in front of me from
my fifth or sixth year, all objects and people that I
knew as time went on, and with whom I had some
relations, all my games, occupations, actions, one
following the other, day after day, year after year,
in one word the picture of my whole past became
present in my sight…"

Among the handful of sympathetic observers
was Carl von Dittmar, who in 1900 wrote about
the Siberian practices in St. Petersburg. He reported, "Mukhomor eaters describe the narcosis as
most beautiful and splendid. The most wonderful
images, such as they never see in their lives oth-
erwise, pass before their eyes and lull them into a
state of the most intense enjoyment…. Among the
numerous persons whom I myself have seen intoxicated in this way, I cannot remember a single one
who was raving or wild. Outwardly the effect was
always thoroughly calming—I might almost say,
comforting. For the most part the people sit smiling
and friendly, mumbling quietly to themselves, and
all their movements are slow and cautious."

These few accounts are about the only information on the effects of Amanita muscaria on Siberian
natives, even though it grew plentifully in Koryak
territory. During the off-season a reindeer would often be exchanged for just one of these mushrooms.
When questioned, natives who had used Amanita
muscaria said repeatedly and emphatically that
they liked it better than alcohol. Eventually alcohol
seems to have supplanted use of this mushroom.

Panther Caps

AFTER WASSON STIRRED INTEREST in Fly Agaric, it didn't
take long for people to notice that its close relative,
the Panther Cap, also has psychoactive effects. The
cap for this species comes in yellowish to grayish
brown. Like the Fly Agaric, it is usually covered by
prominent white warts, which are the remains of
the veil that encloses it when young.

The stem of Amanita pantherina bears a large,
lacy ring, or collar. Appearing at its base is this species' most important distinguishing feature—two or
three layers or hoops of tissue attached to the stem,
which are other rem-
nants of the veil, left
over from when the
stem expands. The
Panther Cap, like the
Fly Agaric, can be
found throughout
woodlands—under
trees or near stumps.
The spores of both are
white, so a spore print
should be taken on
dark paper.

Presence of God

IN MANY OF HIS ANTHROPOLOGICAL CONJECTURES about
mushrooms, Wasson has applied the question of
Fly Agaric and other mushrooms to the origins
of religious feeling. "I suggest to you that, as our
most primitive ancestors foraged for their food,

they must have come upon our psychotropic mushrooms, or perhaps other plants possessing the same property, and eaten them, and known the miracle of awe in the presence of God. This discovery must have been made on many occasions, far apart in time and space. It must have been a mighty springboard for primitive man's imagination."

In the Vedic hymns, Aldous Huxley wrote that "We are told that the drinkers of soma were blessed in many ways. Their bodies were strengthened, their hearts were filled with courage, joy, and enthusiasm, their minds were enlightened, and in an immediate experience of eternal life they received the assurance of their immortality."

The hymns of this first book of the Vedas undoubtedly vibrate with ecstasy. Users are exhalted. "We have drunk the Soma," they say at their height, "we are become Immortals." In Book IX,

John W. Allen

A slug slowly becomes intoxicated by the Amanita mushroom, as do reindeers and squirrels that love to eat them and frolic playfully in the forests and meadows. This snail, in a state of bliss, was unable to leave the 'shroom and slowly dried up, as did the 'shroom.

trip on Fly Agaric he had to re-evaluate Wasson's SOMA. His feeling was that ingestion of Amanita muscaria was not better than Cannabis, opium, Datura, or betel, all of which were already known in India by the time of the Rig-Veda.

Such objections deserve mention, because results for individual use are variable. Some users may get the exaltation described in the early accounts, like those from Joseph Kopéc and Johann Georgi. Others have experienced little more than sedative or dissociative effects.

19
Drying Amanitas

MPRESSED BY REPORTS about Siberian use of Fly Agaric and feeling this mushroom is in fact the Soma of the Rig-Veda, Wasson experimented on himself and colleagues. "In 1965 and again in 1966 we tried out the fly-agarics repeatedly on ourselves. The results were disappointing. We ate them raw, on empty stomachs. We drank the juice, on empty stomachs. We mixed the juice with milk, and drank the mixture, always on empty stomachs. We felt nauseated and some of us threw up. We felt disposed to sleep, and fell into a deep slumber from which shouts could not rouse us, lying like logs, not snoring, dead to the outside world. When in this state I once had vivid dreams, but nothing like what happened when I took the Psilocybe mushrooms in Mexico, where I did not sleep at all.

"In our experiments at Sugadaira there was one occasion that differed from the others, one that could be called successful. Rokuya Imazeki took his mushrooms with mizo shiru, the delectable soup that the Japanese usually serve with breakfast, and he toasted his mushrooms caps on a fork before an open fire. When he rose from the sleep that came from the mushrooms, he was in full elation. For

three hours he could not help but speak; he was a
compulsive speaker. The purport of his remarks
was that this was nothing like the alcoholic state;
it was infinitely better, beyond all comparison. We
did not know at the time why, on this single occasion, our friend Imazeki was affected this way...."

Drying Effects

SOON AFTER, WASSON NOTICED with great interest that
the Koryaks had told Nikolai Sljunin, who wrote a
two-volume natural history published in St. Petersburg in 1900, that fresh Fly Agaric was poisonous—
and they refrained from eating the mushrooms until they were dried, either by the sun or over a fire.
Wasson linked this with a comment in an heroic
hymn of the Vogul people, in which the hero, "the
two-belted one," addressing his wife, says, "Woman, bring me in my three sun-dried fly-agarics!"

Drying Affects Strength

The drying of this mushroom tremen-
dously affects the strength and nature of
the mental experience. Decarboxylation of
ibotenic acid, the main psychoactive, into
muscimol multiplies activity by a factor of
five or six while reducing the undesirable
side-effects of fresh Amanita muscaria.
Those that are fresh may be dangerous or
not satisfactorily "bemushrooming."

The Rig-Veda had not prepared me for
the drying. I had known of course that the
Soma plants were mixed with water before
being pounded with the pressing stones,
but I had supposed that this was to freshen

up the plants so that they would be capable of yielding juice when pressed. The desiccation, I thought, was an inevitable consequence of bringing the mushrooms from afar and keeping them on hand. There was nothing to tell me that desiccation was a sine qua non of the Soma rite. The reader may think that I should have familiarized myself with the Siberian practice before going to Japan. I agree. Imazeki, who by chance toasted his caps on one occasion before eating them, alone had satisfactory results, insistently declaring that this was nothing like alcohol; that this was far superior, in fact in a different world. Alone among us all, he had known amrta, the ambrosia of the Immortals....

—R. Gordon Wasson

Accidental vs Deliberate Use

ANDREW WEIL WRITES IN The Marriage of the Sun and Moon that he was interested in tracking down instances of Amanita pantherina ingestion.

Accidential Ingestion

They were foraging for edible species and made a mistake. Thinking the panther was some innocuous edible, they took it home, cooked it, and ate it. This mushroom produces an intoxication of rapid onset. Within 15 to 30 minutes, it made all of these people feel very peculiar....

When they began to feel peculiar, all of them decided they had eaten a poisonous

species and were about to die. One woman
first called her lawyer to change an item in
her will, then summoned an ambulance.
All of them got sick. All lost consciousness
for varying periods of time, from a few
minutes to a half hour. All were taken to
emergency wards of hospitals, where they
uniformly received incorrect medical treatment: large doses of atropine that made
their conditions worse. They were admitted
to medical wards and discharged in 36 to
48 hours, since it is the nature of the intoxication to subside quickly, usually within
12 hours. Most of these victims said they
would never eat mushrooms again. One
man said he could not look at mushrooms
in the store for months afterward. When
told some people ate the mushroom for
fun, they shook their heads in disbelief.

—Andrew Weil

Weil found the second kind of Amanita pantherina use among people who had already had extensive experience with psychedelics and "believed
that nature provides us with all sorts of natural
highs just waiting to be picked in the woods....
When these people felt the rapid effects of Amanita
pantherina, they welcomed them as signs that the
mushroom was really working. None of them got
sick. (A few mentioned transient nausea but did not
regard it as important.) None of them felt it necessary to summon help. All of them liked the experience and most said they intended to repeat it. Some
had already eaten the panther a number of times."

Weil presented this information to groups of physicians who tried hard to come up with some simple, materialistic explanation for the response-difference in the two cases. Weil said that the question they always ask is: "Might there have been a dose difference?...There was a dose difference—the people who ate the panther deliberately ate more of it than the people who ate it accidentally."

Psychoactivity

WHEN AMANITA MUSCARIA WAS FIRST examined in 1869 to determine its psychoactive components, a compound called muscarine was isolated. From then until nearly a century later, this compound was believed to be the cause of Fly Agaric's mental effects. Later research have shown that muscarine alone raises quite different responses than the mushroom. In fact, this molecule is present in only trace amounts—0.0002-0.0003% of the fresh plant.

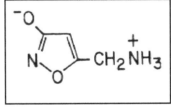

Muscimol
(3-hydroxy-5-amino
methyl isoxazole)

Ibotenic Acid
(amino-3-hydroxy-
5-isoxazole acetic acid)

Since the mushroom's mental effects are undisputedly strong, other guesses were made, notably atropine and bufotenine. Such ideas have now been discarded. Modern research into this question began in 1967 with the work in Zurich of the

chemist C.H. Eugster and the
pharmacologist P.G. Wasser,
who discovered that the main
psychoactives are ibotenic
acid, muscimol and musca-
zone, aided possibly by a few
other constituents.

Amanita verna

Ibotenic acid, considered
somewhat toxic, amounts to
about 0.03-0.1% of the fresh Fly
Agaric mushroom. The noticeabie
differences in the impact of fresh and
dried Fly Agaric mushrooms probably results from
the transformation of ibotenic acid during the drying process into the more potent and quite stable
muscimol. Taken orally, muscimol displays activity
at 10-15 mg. Ibotenic acid is active above 90 mg.
The other psychic contributor, muscazone, seems to
have considerably less effect.

Fly Agaric and Panther Cap specimens should
be dried before ingestion. This is best done on a
drying rack or by suspending them on a wire and
leaving them in the sunlight. Properly dried, these
mushrooms retain their psychoactivity for at least a
year without marked diminution in potency.

The Urine High

THE PSYCHOACTIVE PRINCIPLES in Amanita muscaria
pass through the human organism in such a way
that they are still psychoactive when they emerge
in urine. The resulting sequential inebriation quite
fascinated the explorers and travelers who first witnessed Fly Agaric use in Siberia. The custom there
was delicately described by the English novelist
Oliver Goldsmith in 1762. "The poorer sort, who

love mushrooms to distraction as well as the rich, but cannot afford it at first hand, post themselves on these occasions around the huts of the rich and watch the ladies and gentlemen as they come down to pass their liquor, and hold a wooden bowl to catch the delicious fluid, very little altered by filtration, being still strongly tinctured with the intoxicating quality. Of this they drink with the utmost satisfaction and thus they get as drunk and as jovial as their betters."

Physical Effects

REPORTS ON THE EXPERIENCE of Fly Agaric and Panther Cap mushrooms are fairly rare but seem to indicate that somatic sensations vary considerably, a result of dosage differences, the time of year when they are picked—effects seem to decline at the end of the season—and whether they have been dried.

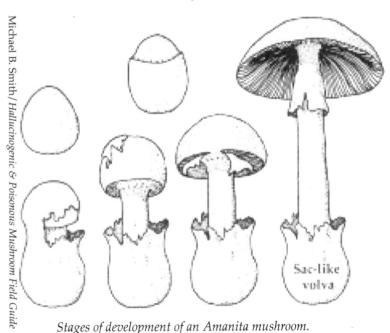

Michael B. Smith / *Hallucinogenic & Poisonous Mushroom Field Guide*

Sac-like vulva

Stages of development of an Amanita mushroom.

According to The First Book of Sacraments, written by the Church of the Tree of Life, the effects of Fly Agaric include: "…Twitching, dizziness, and possibly nausea about half an hour after ingestion followed soon by numbness of the feet. At this point a person will frequently go into a half-sleep state for about two hours. He may experience colored visions and be aware of sounds around him, but it is usually impossible to rouse him. After this a good-humored euphoria may develop with a light-footed feeling and perhaps an urge to dance. At this time a person often becomes capable of greater than normal feats of strength. Next hallucinations may occur. Objects may appear larger than they are. Sometimes a person may feel compelled to reveal harbored feelings. The post-sleep stage may last three or four hours."

Poisonous Amanitas

EATING AMANITAS CAN BE LETHAL. The Amanita muscaria and the Amanita pantherina belong to a genus that is credited with about 95 percent of all deaths resulting from mushroom poisoning. Amanita pantherina—commonly known as the Panther Cap—contains the same psychoactive ingredients, but usually in greater concentrations. Three of their relatives—Amanita virosa, called the Destroying Angel, Amanita verna, and especially Amanita

Amanita phalloides or Death Cap

Mushroom Collecting for Beginners Canada Dept of Ag.

Amanita pantherina,
the Panther Cap

phalloides, or Death Cap—contain lethal toxins that affect the liver and kidneys.

The victims usually do not feel these toxins until about two days after they have eaten the mushrooms, by which time pumping the stomach and other medical measures seldom prove effective. Modern medicine has had limited success counteracting these virulent poisons by use of dialysis machines and blood transfusions. Because Amanita muscaria and Amanita pantherina are similar in appearance to their lethal relatives, it is recommended that one never eat any Amanita that is all white.

Amanita muscaria's alternate name—Fly Agaric—is said to come from the belief that flies can be killed by means of this mushroom. When Wasson tried the experiment, the flies became temporarily stupified but recovered.

Although half of the references pronounce this species deadly, Wasson claims that there isn't a single firsthand account of lethal poisoning by Amanita muscaria. In fact, he asserts that "most trustworthy observers" testify that, "properly dried, it has no bad effects."

After witnessing a considerable number of Fly Agaric and Panther Cap experiences, Jonathan Ott agrees, but he urges potential users to start with no more than 1/4-1/2 cup of chopped or sautéed material.

Jonathan Ott warns that "The genus Amanita possesses at least five species which are potentially lethal. Unless you are very skilled in identification of the Amanita species, do not eat an Amanita that is all white. Caution should also be exercised with regard to dosage. These mushrooms are powerful. The effective dose range may be narrow. If it is exceeded, even by a small amount, a dissociative experience may result, even a comatose state or an inability to function. Of course, there are many who desire this type of effect; no doubt it would be alarming to others. There are many unanswered questions concerning the toxicity of these mushrooms. It has been suggested, and there is some evidence to support this, that the toxicity may vary according to location and season."

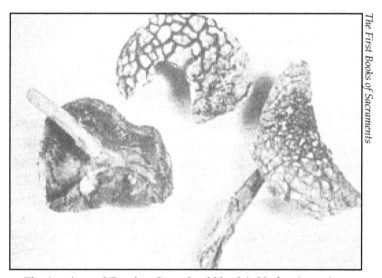

The First Books of Sacraments

Fly Agarics and Panther Caps should be dried before ingesting.

Dosage

PROPER DOSAGE DEPENDS ON MANY VARIABLES. Potency declines at the end of the season. There is much variation between specimens grown in different locales. Reid Kaplan, Wasson's chief colleague in the study of native North American practices, illustrates in his own case how variable this mushroom can be. He failed to feel effects after repeated tries on an empty stomach, with lime, as an enema, etc. These efforts were without any success until he was guided in its use by Keewaydinoquay.

Reports on Siberian tribesmen suggest that they usually took three—one large and two small sun-dried mushrooms, often with reindeer milk or bogberry juice, which is similar to blueberry juice. However, the Church of the Tree of Life cautions potential users that "Siberian tribesmen have a far more robust constitution than most of us." They suggest that no one take any more than a single, modest-sized mushroom—at least until its effects are clearly demonstrated. According to Jonathan Ott, all parts of the Fly Agaric are psychoactive, although the skin of the cap is the most potent part.

Mental Stillness

ACCORDING TO JONATHAN OTT, "After oral ingestion, the full effects will begin in about 90 minutes. For me, these are characterized by wavy motion in the visual field, an "alive" quality to inanimate objects, auditory hallucinations, and a sense of great mental stillness and clarity. The effects are distinctly different from psilocybin, LSD, or mescaline, and may last up to eight hours. Side effects often include nausea, slight loss of balance and coordination, and ligion and Alchemistry (originally titled Strange Fruit). It is about Amanita muscaria and furthers R.G. Wasson's speculations that this mushroom was critical in precipitating "The Origin of Religion." Robert Anton Wilson put the matter squarely:

> If, as it seems, this mushroom was the reality behind several of the legendary drugs of early European and Asian mythology,, then we might conclude that this mushroom played a larger role in religious history than any other single factor.

There are associations in Heinrich's book that support this view. As he told me, in some ten

years of his book's existence, no one has critically objected to its central hypothesis!

Glossary

Agaricales

The large botanical order of gill-bearing fungi to which psilocybians belong.

Basidia

The tip of cells near the mushroom gills where psilocybian mushrooms begin as microscopic spores.

Basidiomycetes

A large class of higher fungi bearing spores on a basidium, having septate hyphae (includes rusts, smuts, mushrooms and puffballs).

Carpophores

The gilled "fruiting body" of a mushroom, the stalk above ground, that is its sexual, "flowering" part.

Casing

The procedure of covering a mushroom's mycelium in the vegetative stage with about half an inch of slightly alkaline soil that greatly increases yield by causing it to "fruit" repeatedly in "flushes" that appear periodically.

Curandera

A "wise woman," such as Maria Sabina, who is somewhat like a shaman in other traditions, and conducts a velada.

Flush

Another round of mushroom growth, generally coming up about a week apart, after the original specimen has been harvested and then its soil has been" cased."

Hyphae

Thin, threadlike cells that mass together and spread underground to become a mushroom's mycelium which corresponds to the roots of a green plant.

Ibotenic Acid

The main psychoactive in fresh Fly Agaric mushrooms, which can be felt at about 90 mg. When these mushrooms are dried, they are transformed into muscimol, which is about five times as strong and quite stable.

Indole Ring

A crystalline component of eight carbon atoms, seven hydrogen atoms, and one of nitrogen, that forms the "nucleus" of most psychedelic compounds.

Indole Tryptamines

Crystaline amides, with ten carbon atoms, twelve hydrogen atoms and two nitrogens, that are the building blocks for most psychedelics.

Muscarine

Thought at one time to be the psychoactive component of Amanite muscaria, but now known to be false.

Muscazone

The third most active contributor to fresh Fly Agaric effects.

Muscimol

The second most active contributor to the psychoactivity of Amanite muscaria and Amanite pantherina. When dried properly ibotenic acid is altered into muscimol, resulting in enhanced psychoactivity.

Mushroom Stones

Artistic productions that may have been the symbols of a "mushroom religion." Found mainly in the Guatemalan highlands, in El Salvador, and in southeastern Mexico.

Mycelium

The underground massing of interwoven filamentous hyphae of a mushroom that could be thought of as eventually becoming the "roots" of the plant. When it "fruits" the sexual part of the organism appears as what we ordinarily think of as a "mushroom."

Mychophiles
People who "love" mushrooms.

Mychophobes
People who fear mushrooms and toadstools.

Mycorrhizal Relationship
The need for Fly Agarics and Panther Caps to form a symbiotic relationship with the root cells of a handful of specific trees if they are to grow.

Psilocin
The secondary activating substance in psilocybian mushrooms. Its presence generally means that the strength of such a mushroom will deteriorate fairly rapidly.

Psilocybe
The genus to which most psychoactive mushrooms belong, based on microscopic characteristics, particularly of the spore.

Psilocybin
The major psychedelic agent in psilocybian mushrooms, which by itself is quite a stable compound.

Psycholytic Therapy
Low-dose use of a psychedelic to elicit buried childhood memories, to aid in building rapport between therapist and client, and to enhance verbalization. Practiced mainly in Europe.

Spores
Microscopic entities that grow on the tip of cells near a mushroom's gills and are its route to reproduction.

Stropharia
An alternative name for the genus of most psychedelic mushrooms, based on the macroscopic features of a psilocybian.

Teonanacatl
A Spanish word from the 16th Century meaning wondrous, awesome or divine mushroom.

Tryptamines
A crystalline amine derived from tryptophan, which is widely distributed in proteins and is essential to animal life, that is characteristic of most psychedelic compounds.

Related Books

Allen, J.W. and J. Gartz. Magic Mushrooms in Some Third World Countries, Psilly Publications, 1997.

Arora, D. All That the Rain Promises, and More: A Hip Pocket Guide to Western Mushrooms, Ten Speed Press, 1991.

Cameron, E. and R. Henneberger. The Wonderful Flight to the Mushroom Planet, Little Brown & Co., 1988.

Czarnecki, J. and L.B. Wallach. A Cook's Book of Mushrooms: With 100 Recipes for Common and Uncommon Varieties, Artisan Sales, 1995.

Enos, L. A Key to the North American Psilocybe Mushroom, Youniverse Press, 1973.

Estrada, A. Maria Sabina, Her Life, Her Chants: An Autobiography, Ross-Erikson, 1976.

Friedman, S.A. Celebrating the Wild Mushroom, Dodd, Mead and Co, 1987.

Gartz, J. Magic Mushrooms Around the World, Luna Information Services, 1997.

Gottlieb, A. Peyote and Other Psychoactive Cacti, Ronin Publishing, 1997.

Gottlieb, A. Psilocybin Production, Ronin Publishing, 1997.

Grubber, H. Growing the Hallucinogens: How to Cultivate and Harvest Legal Psychoactive Plants, Ronin Publishing, 1992.

Harris, Bob. Growing Wild Mushrooms: A Complete Guide to Cultivating Edible and Hallucinogenic Mushrooms, Ronin Publishing, 2003.

Laesse, T., et al. Eyewitness Handbook: Mushrooms, DK Pub. Merchandise, 1998.

Lincoff, G. and D.H. Mitchel (eds.). Toxic and Hallucinogenic Mushrooms, Van Nostrand Reinhold, 1980.

Lonik, L. Basically Morels: Mushroom Hunting, Cooking, Lore and Advice, Stackpole Books, 2001.

McKenna, T. The Archaic Revival, Harper SanFrancisco, 1992.

McKenna, T. Food of the Gods: The Search for the Original Tree of Life, Bantam Books, 1993.

McKnight, K.H. and V.B. McKnight. A Field Guide to Mushrooms: North America, Houghton Mifflin Co., 1998.

Menser, Gary P. and Gery P. Menser. Hallucinogenic and Poisonous Mushroom: Field Guide, Ronin Publishing, 1996.

Miller, M. and C. Nelson. Chanterelle: The Rocky Mountain Mushroom Book, Johnson Publishing Co., 1988.

Oss, O.T. and O.N. Oeric. Psilocybin: Magic Mushroom Growers Guide: A Handbook for Psilocybin Enthusiasts, Quick American Archives, 1992.

Ott, J. Hallucinogenic Plants of North America, Wingbow Press, 1979.

Ott, J. Pharmacotheon: Entheogenic Drugs, Their Plant Sources and History, Natural Products Co., 1996.

Ott, J. and J. Bigwood (eds.). Teonanacatl: Hallucinogenic Mushrooms of North America, Madrona Press, 1978.

Pendell, D. and G. Snyder. Pharmako/Poeia: Plants, Powers, Poisons, and Herbcraft, Mercury House, 1995.

Reidlinger, T.J. (ed.). The Sacred Mushroom Seeker: Essays for R. Gordon Wasson, Dioscordies Press, 1990.

Rumack, B. and E. Saltzman (eds.). Mushroom Poisoning Diagnosis and Treatment, CRC Press, 1978.

Sanford, J. In Search of the Magic Mushroom, Clarkson N. Porter, 1966.

Schultes, R.E. Hallucinogenic Plants, Golden Press, 1978.

Schultes, R.E., A. Hofmann, and C. Ratsch. Plants of the Gods: Their Sacred, Healing, and Hallucinogenic Powers, Healing Arts Press, 2001.

Snow, O. LSD-25 and Tryptamine Synthesis: Overview and Reference Guide for Professionals, Thoth Press, 1998.

Stafford, P. Psychedelics Encyclopedia, Ronin Publishing, 1992.

Stamets, P. A Field Guide to Psilocybin Mushrooms of the World, Ten Speed Press, 1996.

Stamets, P. Growing Gourmet and Medicinal Mushrooms, Ten Speed Press, 2000.

Stamets, P. Psilocybe Mushrooms and Their Allies, Homestead Book Co, 1978.

Stamets, P. Psilocybin Mushrooms of the World: An Identification Guide, Ten Speed Press, 1996.

Stamets, P. and J.S. Chilton. The Mushroom Cultivator: A Practical Guide to Growing Mushrooms at Home, Agarikon Press, 1983.

Turner, N.J. and A.F. Szczawinski. Common Poisonous Plants and Mushrooms of North America, Timber Press, 1991.

Wasson, R.G. The Wondrous Mushroom: Mycolatry in Mesoamerica, McGraw-Hill Book Co., 1980.

Weil, Andrew. Marriage of the Sun and Moon, Houghton Mifflin, 1980.

Author Bio

Peter Stafford (right), with Stanley Krippner (middle)
and John Lilly (left).

PETER STAFFORD HAS WRITTEN ABOUT MAGIC MUSHROOMS
and psychedelics since the 1960s. He recorded
events as they happened. "The Stafford Collection"
of material on psychedelics has been deposited in
the Columbia University library. He is best known
for his classic book, Pychedelics Encyclopedia.
Peter lives in Santa Cruz, California. His website
can be found by searching the net for "Psychedelics
101."

Index

TEONANÁCATL
A Bibliography of Entheogenic Fungi

John W. Allen & Jochen Gartz, Ph.D.

Forward: Jonathan Ott

$39.95

Available at
www.mushroomjohn.com

The divine mushrooms have been used by native healers, shaman, and sorcerers for over 3000 years. Archeological evidence suggests that their use actually dates as far back as 9000 B.C. There are more than 214 entheogenic mushrooms of which 186 contain the alkaloids psilocine and/or psilocybine. Many are still hidden from world view. References to their existence is obscure to botanists and historians.

Allen and Gartz have brought together more than 2400 references on entheogenic fungi of which 1600 are annotated and a cross-referenced index provides direct links to more than 8000 author/date citations. It includes over 700 color photographs representing the history of hallucinogenic mushrooms and their worldwide use. This CD is a must-have for scholars and students of mycology—especially those interested in entheogenic fungi.

John W. Allen, who contributed numerous photographs to Magic Mushrooms and created the image on the front cover, is an amateur ethnomycologist who has studied, photographed and collected fungi since the 1970s in the Southeast and Northwest of the United States, Australia, Southeast Asia and Europe. He has authored 9 books. Jochen Gartz, Ph.D. is a German biochemist and mycologist who has authored more than 50 published scientific papers on the chemistry and cultivation of entheogenic mushrooms.

John W. Allen can be contacted at mjshroomer1@yahoo.com

Printed in the USA
CPSIA information can be obtained
at www.ICGtesting.com
JSHW041935140824
68134JS00013B/125